PRACTICAL BLACKSMITHING

A COLLECTION OF ARTICLES CONTRIBUTED AT DIFFERENT TIMES BY SKILLED WORKMEN TO THE COLUMNS OF "THE BLACKSMITH AND WHEELWRIGHT" AND COVERING NEARLY THE WHOLE RANGE OF BLACKSMITHING FROM THE SIMPLEST JOB OF WORK TO SOME OF THE MOST COMPLEX FORGINGS

VOLUME II

Compiled and edited by

M. T. RICHARDSON

ILLUSTRATED

Published by Left of Brain Books

Copyright © 2022 Left of Brain Books

ISBN 978-1-396-32138-2

First Edition

All rights reserved. No part of this publication may be reproduced, distributed, or transmitted in any form or by any means, including photocopying, recording, or other electronic or mechanical methods, without the prior written permission of the publisher, except in the case of brief quotations embodied in critical reviews and certain other noncommercial uses permitted by copyright law. Left of Brain Books is a division of Left of Brain Onboarding Pty Ltd.

Table of Contents

PREFACE.	1
CHAPTER I. IRON AND STEEL.	2
Their Antiquity and Great Usefulness.	2
The Strength of Wrought Iron and Steel.	5
The Rotting and Crystallization of Iron.	7
Steel and Iron.	8
Modern French and English Wrought Ironwork.	10
Upsetting Steel and Iron.	13
Heating Steel in the Blacksmith's Fire.	18
Testing Iron and Steel.	21
The Treatment of Steels.	26
Hints About the Treatment of Steel.	27
On the Working of Steel.	28
Hardening Steel.	29
To Select Good Tool Steel.	34
Different Kinds of Steel.	35
Restoring Burnt Steel.	35
Cold Hammering Iron.	35
CHAPTER II. BOLT AND RIVET CLIPPERS.	38
A Bolt and Rivet Clipper.	38
Cut Nippers.	39
Bolt Clipper.	41
A New Bolt Clipper.	42
A Handy Bolt Cutter.	43
Making a Bolt Clipper.	45
Tool for Cutting Rivets.	47
Rivet Cutter.	48

Tools for Making Rivets—Pipe Tongs.	49
A Tool for Making Rivets.	50
Making a Bolt Clipper.	51
How to Make a Bolt and Rivet Cutter.	52

CHAPTER III. CHISELS. 53

The Chisel and Chisel-Shaped Tools.	53
Chipping and Cold Chisels.	67
How to Make Cold Chisels.	71
Forging Cold Chisels.	72

CHAPTER IV. DRILLS AND DRILLING. 73

Making a Drill Press.	73
Drilling in the Blacksmith Shop.	74
A Simple Drill Press.	77
Making a Small Drill.	78
To Drill a Chilled Mold-Board.	79
Holding Long Bars in Drilling.	79
Drilling Glass.	80
Straightening Shafts or Screws—A Remedy for Dull and Squeaking Drills.	80
A Chinese Drill.	81
A Drill and Countersink Combined.	82
A Handy Drill.	83
A Home-Made Drill.	85
Making and Tempering Stone Drills.	87
Some Hints About Drills.	88
Drifts and Driftings.	91

CHAPTER V. FULLERING AND SWAGING. 95

The Principles of Fullering.	95
About Swages.	99
Rules for Swaging.	105

A Stand for a Swage-Block.	107
CHAPTER VI. MISCELLANEOUS TOOLS.	**109**
The Principles on Which Edge Tools Operate.	109
Hints on The Care of Tools.	109
Names of Tools and Their Pronunciation.	111
Tongs for Bolt-Making.	114
Home-Made Fan for Blacksmith's Use.	115
Making a Pair of Pinchers.	118
A Handy Tool for Holding Iron and Turning Nuts.	119
A Handy Tool to Hold Countersunk Bolts.	119
Making a Pair of Clinching Tongs.	121
Tongs for Holding Slip Lays.	122
CHAPTER VII. MISCELLANEOUS TOOLS. CONTINUED.	**124**
Mending a Vise.	124
A Cheap Reamer.	125
Shapes of Lathe Tools.	125
Useful Attachment to Screw Stock Dies.	130
Wear of Screw-Threading Tools.	131
Tool for Wagon Clips.	133
A Handy Tool.	135
False Vise-Jaws for Holding Rods, etc.	136
Making Spring Clips With Round Shanks and Half-Round Top.	137
Handy Tool for Marking Joints.	139
Tools for Holding Bolts in a Vise.	140
A Tool for Making Singletree Clips.	142
Tool for Making Dash Heels.	143
Mending Augers and Other Tools.	146
An Attachment to a Monkey Wrench.	147
A Handy Tool for Finishing Seat Rails, etc.	148

A Tool for Pulling Yokes on Clips.	149
Making a Candle Holder.	149
Making a Bolt Trimmer.	151
A Labor-Saving Tool.	151
Making a Spike-Bar.	152
How to Make a Tony Square.	152
An Easy Bolt Clipper.	153
A Tool for Pulling on Felloes.	154
How to Make a Handy Hardy.	154
A Handy Clincher.	155
A Bolt Holder.	155
Making a Cant-Hook.	156
Making a Cant-Hook.	156
Making Screw Boxes for Cutting Out Wooden Screws.	158
Mending a Square.	159
Stand for Carriage Bolts.	160
An Improved Crane and Swage Block.	162
A Cheap Crane for Blacksmiths.	163
Repairing an Auger.	164
A Clamp for Holding Countersunk Boltheads.	165
A Handy Machine for a Blacksmith.	165
A Clamp for Framework.	166
A Tool for Holding Bolts.	167
A Hint About Callipers.	167
Vise Attachment.	168
Bolt Set	169
A Home-Made Lathe.	169
CHAPTER VIII. BLACKSMITHS' SHEARS.	**179**
Shear for Cutting Round and Square Rods.	179

Cheap Shears for Blacksmiths' Use.	180
Blacksmiths' Shears.	182
Shears for the Anvil.	183
CHAPTER IX. EMERY WHEELS AND GRINDSTONES.	184
Emery Wheels.	184
Making an Emery Wheel.	184
How to Make Small Polishing or Grinding Wheels.	185
Making an Emery Wheel.	186
Something About Grindstones and Grinding Tools.	187
Hanging a Grindstone.	189
Device for Fastening a Grindstone.	189
Mounting a Grindstone.	190
How to Make a Polishing Machine.	192

PREFACE.

In Vol. I. the editor of this work gave a brief account of the early history of blacksmithing, so far as known, and described a few ancient and many modern tools.

Numerous plans of shops were given with best methods of building chimneys and constructing forges.

This volume opens with a brief treatise on the early history of iron and steel. Artistic iron work is then considered, and the tests employed to show the strength of iron are given.

It was the original intention to compress all the material relating to tools in Vol. I., but this was found to be impracticable without largely increasing the size of the volume, and possibly the price as well. It was deemed best therefore to devote so much of Vol. II., as might be necessary, to the further consideration of tools, and the reader will doubt-less agree with us that the space has not been wasted.

Vol. III. will be devoted mainly to the consideration of jobs of work.

The Editor

CHAPTER I.

IRON AND STEEL.

THEIR ANTIQUITY AND GREAT USEFULNESS.

All mechanics, irrespective of the trades which they follow, have so much to do with iron in its various forms, either by working it or by using tools and instruments in the construction of which it forms an important part, that a brief consideration of the material, both retrospective and present, cannot fail to be of interest. To trace the development of iron from its earliest known existence to the present, and to glance at its use as a material of construction, that, unlike all others can rarely be dispensed with in favor of a substitute equally desirable, cannot fail to be of the greatest interest.

In the description of the building of Solomon's Temple there is no specific statement that iron was employed, although by inference it is understood that this material must have been used in the tools of the workmen, if for no other purpose. It is not to be forgotten that the record says there was neither hammer nor axe, nor tool of iron heard in the house while it was being built. A plausible construction to place upon this assertion is that the parts were fashioned and fitted together at distant places, and were joined noiselessly in completing the structure. This view of the case serves to point out the excellence of the skill of that day, for, however great the care that is exercised at the present time by mechanics, few buildings are put up in which the sound of tools in the shaping of the various parts, after they have been sent to the building for putting up, might be dispensed with.

King David, it is said, had in his collection of materials "iron in abundance for the doors of the gates and for the joinings." Other allusions to iron are to be found in the Old Testament Scriptures. Some passages are of figurative character, where iron is the emblem of hardness, strength and power. Others are descriptive, and indicate its uses in those early times.

The manufacture of iron existed in India from very remote antiquity, though carried on in a very primitive manner. Vast accumulations of slag and cinder are found spread over large areas in various districts, and the manufacture of iron is still carried on with little change from the ancient process. In one of the temples near Delhi there is a wrought-iron pillar sixty feet in length, which dates as far back as the fourth century of the present era. It is only of late years that the production of a shaft of these dimensions has been possible to the present race of iron-workers.

The extent to which iron was employed by the ancient Egyptians is a problem difficult of solution. When the enormous labor expended upon the masonry and sculpture in the hardest granite, and the beautiful surface and high finish generally displayed in the architectural works of that country are considered, it seems difficult to imagine that tools inferior to the hardest steel could have produced the result, yet the evidence is exceedingly slight that anything of the kind was employed. Bronze is sometimes found in ancient tombs in that country in a variety of forms, but iron is almost entirely wanting. Iron mines have been discovered in Upper Egypt, and the remains of iron-works have been found recently near Mount Sinai. An iron plate was found in one of the pyramids, and a sickle in one of the tombs at Thebes.

The use of iron by the Romans was of comparatively late introduction. The fine specular iron of the Isle of Elba had been smelted by the Etruscans from an early date, but it does not seem to have been extensively used in Italy. It was not until the time of the second Punic War that the Romans, extending their conquests beyond their narrow original seat, obtained supplies of iron and steel from Spain, and discarded their bronze weapons for the harder and keener metal. Iron was little used by any of the ancient nations in building construction. When Virgil describes the splendors of Dido's rising city, no mention is made of iron in any form. Stone, with the addition of bronze for plating, are the materials especially alluded to.

In the New Testament mention is made of an iron gate which would seem to indicate that at that period iron had been brought into use in many forms, more, however, in the way of machinery, armor and weapons, than in building construction. For all constructive purposes bronze was gradually superseded by iron, and during the middle ages was worked with great skill and success. Iron employed at this period was not made by the process of fusion and puddling, but was obtained direct from the ore by roasting with charcoal and working it under the hammer. The metal thus obtained was of excellent quality, and such examples as have come down to us indicate that it was very skilfully manipulated. Armor and weapons attained a high degree of efficiency, and were finished with great taste.

There are very few specimens remaining in anything like perfection of the mediaeval smith's work. Enough fragments, however, are in existence to indicate the extreme beauty of the workmanship of this age. The rich and graceful curves of the work done at this time, together with their lightness and strength, show what capabilities exist in iron when freely treated in accordance with its nature. Probably the grills or screens and the gates of the middle ages exhibited the art of the smith in its greatest perfection. Nothing that has been made in modern times is equal to the specimens which remain of that period. The wonder arises that, with such simple means as were at the command of the mechanics and artists of that day, such wonderful effects could be produced.

During the Romanesque period iron does not seem to have been employed, even in carpentry or masonry. At the end of the twelfth century iron cramps were employed at the Cathedral of Notre Dame, Paris, to connect the stones of the cobbled corners. The oxidation of these cramps in the course of time had the effect of fracturing the stones. Experience in this practice does not seem to have taught wisdom, for, even at the present day, the mistake of using iron in similar positions has frequently led to like results.

We owe to Germany the discovery of the process by which fusible iron could be smelted from the ore. It probably arose from the gradual improvement of the blowing apparatus, by which the old blast bloomeries were transformed into blast furnaces. Cast iron was unknown prior to the middle of the sixteenth century. About 1550 the German system, above alluded to, was introduced into England, where there already existed great facilities in the enormous quantity of scoriae accumulated about the ancient bloomeries and in the abundance of timber for fuel. Progress in the art was so rapid that cast-iron ordnance was an article of export from England early in the seventeenth century. As the art of casting made progress, the art of the smith declined. The cheapness of cast iron and the facility with which it could be manipulated, led to its extensive use in every department of life.

In 1735, the problem of smelting iron with pit coal was successfully solved, but it was not until within a very recent period that the advantages of iron on any great scale developed themselves. Down to the commencement of the present century the casting of iron pipe was so difficult and costly an operation, that in schemes for the supply of water to towns, wooden pipes were adopted for the mains. In 1777 the first experiment was made with iron as a material in bridge building. At present scarcely a bridge of any importance is constructed of other material.

Having thus glanced hastily at the progress of manufacture and the use of iron from the earliest historical periods to the present time, the inquiry comes up, what is to be its influence in the future? That it will contribute materially to aid man's power over the elements of nature is certain, but the moral results which are likely to follow lie beyond our province. All true designs arise out of construction. Every style which has attained any eminence owes its effect to the adoption of its essential parts as sources of beauty rather than an attempt to conceal them. The use of iron in any construction or design is a source of power and effect, put into the hands of the architect for good or for evil.

THE STRENGTH OF WROUGHT IRON AND STEEL.

There is something very interesting, but not altogether as yet understood, in the behavior and strength of iron and steel when loaded.

It is all very well to institute certain tests to find the number of pounds it requires to break a piece having a sectional area of one square inch, and from this pronounce what is the strength of the iron; because, with our present knowledge and appliances, it is all we can do, and a test of some kind is of course imperative. It is a curious fact, however, that the strength of a piece of iron or steel varies according to the manner in which the load is applied, If the metal receives its load suddenly, it will break under a less weight than if the load comes on slowly and gradually increases; and the difference is not a minute one either, for it is as great as 20 per cent under the two extremes of conditions. One of the most eminent constructing engineers in this country stated not long since, in reply to a question, that he would make as much as

20 per cent difference in the strength of two beams to receive the same load, one to have the load suddenly, and the other to have it gradually applied. From this it is a fair and reasonable deduction that if the load, when applied, caused vibration, the beam would require still greater dimensions to be of equal strength, because vibrations are simply minute movements, and, in the case of horizontal beams, on moving downward increase the pressure of the load.

A short time since some experiments were made to ascertain the strength of iron and steel wire, two specimens of each size of wire being used, one just as the iron came from the mill and the other an annealed specimen.

The wires were suspended vertically, and a certain weight, as say 10 lbs., was hung on them. Then in some cases a ½ lb. weight per day was added, in others 1 lb. per day, in yet others the weights were increased as fast as they could be put on, and in every instance it was found that the breaking strains increased according as the time between the increases of weight was made longer, the amount varying from 10 to 20 per cent. The failure of the boiler plates of the English steamship *Livadia* elicited some interesting facts and strange opinions upon the behavior of low-grade steel. The facts concerning these plates are given below. The boiler was 14 feet 3 inches diameter by 16 feet long. The plates were ¾-inch thick, lap-jointed and treble riveted. The plates were all punched, then slightly heated and bent to shape, afterwards put together, and the rivet holes reamed out to size. While under this treatment one of the plates fell out of the slings on to an iron plate and was cracked right across the rivet holes. Naturally this gave some anxiety, but after the plates were all in the boiler itself, they cracked across the rivet holes in nearly all directions; that is, many of them did.

Investigation was immediately set up, chemically and mechanically, when it appeared, as nearly as could be ascertained, that although the stock was good of which the plates were made, it had not been thoroughly worked under the hammer before rolling.

Dr. Siemens, the inventor of the process which bears his name, asserts that annealing plates, either before or after working (punching), is of no advantage; tending, if anything, to injure rather than benefit the materials. Many practical men, however, hold views in opposition to Dr. Siemens on this question. —*By* Joshua Rose.

THE ROTTING AND CRYSTALLIZATION OF IRON.

I noticed an article lately, in which an iron worker, who claims an experience of fifty years in his trade, says that iron rots as well as crystallizes under strain and jar. The latter part of this statement is correct in degree only. The springs of vehicles deteriorate by use and excessive strain, but not to the extent which the writer of the article I refer to represents, as the springs of thousands of old vehicles will attest, in which not a leaf is broken, although the remainder of the gear is worn out.

And although iron may be crystalline in the fracture it does not lose its tensile strength to any great extent, unless under a very great strain or jar, as in the case of quartz-mill stamp stems, which are lifted and dropped about once a second, are run night and day, including Sundays, and even then will stand several years of this hard usage before breaking. The danger of iron losing its tensile strength is greatly exaggerated in the article in question. If it were not so people would be afraid to go over and under the Brooklyn bridge.

The statement that iron rots is absolutely untrue, and if the crowbars referred to in the article mentioned would not weld readily, and had a bad smell when heated, the odor was from the sulphur and phosphorus which the iron contained, and which was present in the iron when it was made, like much of the first iron in early attempts at iron making by using stone coal. The iron did not absorb, it could not have absorbed sulphur and phosphorus from age or exposure to the atmosphere.

The art of iron making has progressed in spite of all statements to the contrary, and iron is made smelted with stone coal, which is as good as the best Swedish charcoal iron, and even better for some purposes, and the iron makers manufacture any grade of iron to suit price or purpose.

When iron, by reason of long-continued strain or jar, breaks, showing a crystalline fracture, its chemical constituents are still the same as when it was made, and when heated and welded it will resume its fibrous appearance and its original toughness, as I know from much practical experience in welding broken stamp stems and heavy iron axles.

Rolling mills were invented about a hundred years ago, but if the practical wiseacre is in favor of the old system of laboriously pounding out a bar of iron filled with hammer marks by the trip hammer, why that settles the rolling mills of course. —*By* R. R.

STEEL AND IRON.

I have been turning in my mind some of the generally accepted theories about iron and steel, and wondering when the general public will drop the notion that the main distinguishing feature between the two is that one will harden and the other will not.

Does a piece harden? It is steel.

Is it found impossible to harden another piece? It is iron.

Many people go no further than this in deciding the character of pieces under examination, and still there is steel that will not harden which is almost equal to tool steel. Growing out of these wide differences between different steels and irons there is a continuous discussion and much misconception as to real facts. That a piece of Swedes iron of irregular structure, with minute seams, sand streaks and impurities, should contain sufficient carbon to harden would hardly make it valuable for edge tools. Nor would a piece of cast-steel of the most unexceptionable structure be of any great value for the same use, in a commercial sense, if the necessary carbon were lacking. How often is heard the very positive assertion in regard to certain articles which should be made of steel and hardened, "They are nothing but iron," and quite recently there has appeared in trade journals an article on cutlery, in which it is charged that "table knives are made of iron, on account of the greater facility with which iron can be worked."

That cast-iron shears and scissors with chilled edges, and cast-iron hammers and hatchets sandwiched in with malleable iron and steel castings, to take the place of instruments which are generally supposed to be forged from steel and hardened and tempered in the regular way, are to be found on the market is true, but when it comes to goods which are made in the regular way, we need not believe that wrought iron is used to any great extent where steel should be. That a tool is soft does not prove that it might not have been hardened to be one of the best of its kind. That a tool proves to be as brittle as glass, breaking at the very beginning of service, does not prove that the steel of which it was made was of poor quality, for, properly treated, it might have been hardened to be of the very best.

In this matter of deciding as to the merits of steel there is too much of jumping at conclusions, and so the self-constituted judges are continually called upon to reverse their decisions. No decision would be considered to be in order on a matter of law until the evidence was all in, and not till the evidence was laid before a judge and jury would they be asked to render a decision. No more can a man expect to decide off-hand the character of steel, for what may be attributed to poor quality may be due to bad condition caused by unfair treatment, while to know what would have been fair treatment one must know the quality of steel. Much stress is put upon the fact that only certain brands of imported steel are used by some American manufacturers, who tell us that "they can depend upon it every time—well, nearly every time," and that "they don't have to be so particular about heating it. If it is heated a little too hot it won't crack, but will stand to do something. They like a little leeway." These men seem to think that it is just that particular brand of steel which possesses the qualities of safety of which they think so much, and we hear them say, —s steel does this and that, but you cannot do it with —'s steel, it would fly all in pieces treated the same way."

How is it? Does anybody who has studied upon the subject a little, suppose that when a particular grade of steel of any brand has been found to be right for a certain use, while the first bar used of some other brand without reference to the grade has proved to be apparently of no value, that that settles it, or that ground is furnished for saying that the steel from one manufacturer shows certain characteristics which the other does not?

Then, after a certain grade of any brand of steel has been settled upon as right, it will not do to condemn too broadly steel of another brand, which with the same treatment accorded to the favorite steel fails, for it is quite likely that, with the different treatment which this grade of the new brand requires, it might prove equally as good as the "long-tried and only trusted steel," or, if not of the right grade, a grade could be furnished of the new brand fully as good as the best of the old brand, while it is reasonably certain that had a change of grade without change of brand been made the result would have been much the same. Too much weight is put upon a name; and we hear steelworkers lauding the especially good qualities of this or that steel and condemning the bad qualities of others, when the fact is that both the words of praise and blame apply to the grades of steel and their treatment and condition, not to the fact that this or that was made in Pittsburg or Sheffield. The steel maker, whether he will or not, must, and does, make a variety of grades—tempers—of steel; and upon a judicious and honest selection of the right temper for any particular use, and upon just the treatment required—especially in hardening—hang the desired results.

"Jessup steel doesn't do this," says one of its admirers, but a grade of Jessup's can be had which will, without doubt.

Said an enthusiastic admirer of a "special" brand of imported steel to the writer:

"That steel stands to do what no American steel will do at all; on this work-tools from it have stood to work five and even six hours without grinding."

A few months later this same man said, in speaking of the same work:

"The result in using tools made from the pieces of American steel which you sent by mail were simply wonderful; some of them stood to work without grinding two entire days of ten hours!"

With steel of proper temper from reliable makers, there are possibilities of which many men who look upon steel as steel simply, and who judge its quality in advance by the brand it bears, have never dreamed; and of those who pride themselves on being thoroughly American, and still persist in using English steel, what can be thought, except that they have not carefully investigated to learn the merits of American steel?

MODERN FRENCH AND ENGLISH WROUGHT IRONWORK.

Attention has been called at different times to the possibilities in wrought iron work in the art line, and examples of work have occasionally been presented showing what has been done and may be done in this direction.

Figs. 1 and 2 show two very handsome designs, Fig. 1 being a specimen of wrought ironwork from the establishment of M. Baudrit, of Paris. It is original in design and admirable in execution. There is a charming variety in the work, characteristic of the high productions of French artists. The lower portion is solid, as the foundation of the terminal post of a balustrade should be, but it lies on the stairs naturally and elegantly. The upright pillar and handrail are sufficiently massive, while the decorative portion has all the light elegance of a flower.

FIG. 1—WROUGHT IRON BALUSTRADE BY M. BAUDRIT, OF PARIS.

For it is well known that for every difficult job done with English steel there is the equal done daily with American steel. —S. W. Goodyear, *in the Age of Steel*.

FIG. 2—WROUGHT IRON RAILING BY RATCLIFF & TYLER, OF BIRMINGHAM.

In this country our designers are wont to draw work of this kind for execution in cast iron, and so accustomed have we become to casting all ornamental work of a similar character that our blacksmiths scarcely know what it is possible to accomplish with the hammer and anvil. Fig. 2 is not less striking, and is an example of work in good taste for a similar purpose to that shown in the first instance. It is as unlike it, however, in character and execution as the two nations from which these two pieces of work come. Fig. 2 represents a continuous balustrade executed by Messrs. Ratcliff & Tyler, of Birmingham. An oval in the center is very happily arranged panel fashion between the scroll work which serves the purpose of pilasters. The design is neither too ornamental nor is it poor. The connecting links of the work, including the attachments to the stairs, are graceful and effective. This pattern also, if made in this country, would very likely be executed in cast metal, and would lose all those peculiar characteristics that render it attractive, and, as at present, considered an example of true art workmanship. The mechanical ingenuity of our smiths is universally acknowledged, but in artistic taste and in the ability to execute ornamental work they are very much behind those of other nations.

UPSETTING STEEL AND IRON.

I have recently read some things in relation to the upsetting of iron and steel, which are so much at variance with the generally accepted ideas on the subject, and at the same time so flatly contradict what the every-day experiences of many mechanics show to be facts, as to prompt me to offer some testimony.

First, as to the "Upsetting of Iron," for under this heading may be found in a trade journal of recent date a very interesting reference to the "quality of movements of the particles of iron under pressure or percussion"...

"Red-hot iron can be pressed to fill a mold as clearly and exactly as so much wax could be."... "Cold iron can also be molded into form by pressure."... "The heading of rivets, bolts and wood-screw blanks shows some surprising results in the compression of iron; a No. 6 1-inch screw requires a piece of wire slightly more than 1 ½ inches long to form it. Yet the total length of the screw blank headed is just one inch.... Now, it has been proved by experiments with shorter bits of wire that less than five-sixteenths of an inch of the extra eight-sixteenths is required to form the screw head. What becomes of the remaining more than three-sixteenths of an inch in length of an original 1 ½ inches that make the 1-inch screw blank? There can be but one answer —the iron is driven upon itself,... so that 1 1-16 inches of wire are compressed into seven-eighths of an inch in length without increasing the diameter of the wire."

This flatly contradicts the assertions of scientific investigators, who have, after making many exhaustive experiments, concluded that cold working of iron and steel, such as hammering, rolling, drawing, pressing, upsetting etc., do not increase the specific gravity. Is it likely that in the many careful experiments made by the most painstaking of men, the experiments involving the most accurate measurements possible, added to the unquestionable- tests of specific gravity, there has been uniformly a misconception of the real facts, and that experiments made by measuring a blank piece of wire—possibly with a boxwood rule—before heading, and again measuring the length of blank produced by heading the same wire, are to upset this proven fact, that ordinary cold working does not make iron and steel more dense?

Where does the iron go, then? It goes to round up the contour of the die in which the blank is headed, that entire part of the block constituting the body, and to fill the die for the entire length of the body of the blank, through the upsetting process, to a fullness which in solid dies defies the efforts of any but the best of carefully hardened and tempered steel punches, from best of steel, to push out of the dies after heading rivets and screw blanks, making the question of what steel to use for "punching out in solid die heading," one of the most important connected with the business.

In heading a screw-blank or rivet, the first effect produced by the longitudinal pressure applied is to upset the piece of wire for its entire length. The diminution in length will produce an exactly proportionate increase in diameter up to the point when the wire fills the die so tightly as to transfer the most of the upsetting effect of the continued pressure to that part of the wire not encircled by the die, and then comes the heading, which begins by increasing diameter and proportionately decreasing length, the metal being in a measure held by contact with the die and heading punch from lateral expansion. As the pressure continues, it assumes first a pear shape, simply following the not necessarily written law under which all metals under pressure yield in the direction of the least resistance, and soon a shoulder is formed, which, coming in contact with the face of the die, or bottom of countersink, depending upon the shape of die, the direct or entire resistance to the pressure applied has no longer to be supplied by the body of the wire as at first. But still, as the pear-shaped bulb is gradually pressed out of that shape into the shape required in the completed blank or rivet, a time will come when a portion of the superfluous metal back of the body of the blank, or representing the center of the head, can escape more easily in the direction of the pressure, thereby still further increasing the diameter of the body of the blank, than to escape altogether in a lateral direction under the immense pressure required toward the completion of the heading process.

"Without increasing the diameter of the wire," we quoted, it would not be possible, in any commercial sense, to do anything of the kind. Did the writer who stated as a fact that this was the rule realize the improbability of the statement which he virtually makes, i. e., that over 17 ½ per cent of the wire entering into the body of the blank described by him has been lost, if, as he says, the diameter remains unchanged?

I once made some carefully conducted experiments to prove or disprove the truth of the assertion, which I had often made, that cold swaging did increase the specific gravity of steel, but which was denied by those who, having learned by actual experiment that other cold working did not, felt sure that the effect of cold swaging would be the same as that of other methods. My experiments were made with pieces of steel rod, each two feet in length, by reducing them by cold swaging until they were nearly eight feet long, and, having first carefully measured the original length and diameter, the increased length and decreased diameter was to show by measurement whether the specific gravity had been increased by swaging. I have not the figures at hand, but they showed on my side; still, when I presented them to my opponents, they met me with the statement that "there was too little difference to talk about, and, if my measurement had been absolutely correct, I had established nothing further than to appear to show the exception, which proved the rule. Cold working did not increase density; this was a well-known principle." Well, it was not much. The fraction of an inch which showed the difference in length of pieces from what they should have been, had the specific gravity remained unchanged, was represented by a figure in the third or fourth place in decimals, not much like the 0.1875=3-16 of an inch, which the writer from whom we have quoted would have us believe was lost from 1 ½ inches in length, when it is considered that in my experiments there were pieces sixty-four times as long in which lost metal might hide, and still not more than the one-hundredth part as much had hidden in the eight feet as is claimed for 1 ½ inches.

Some facts are hard to swallow unless dressed up to appear reasonable. "Upsetting iron," as applied to heading, cannot be easily upset by anybody who will take the time to compare diameters and specific gravity of wire before heading with those of headed blanks.

"Upsetting steel" is spoken of by some persons, whose information on the subject of steel working is very extensive, as one of the most pernicious of practices, and, by the way the subject is treated, one might suppose that they fully believed that steel was made up of fibres, which must be always worked in one direction. Steel is of a crystalline nature—not fibrous—and such of it as will be ruined by upsetting within reasonable limits, the operation being performed intelligently, is not good, sound steel.

Pursuing the subject of heading, I can testify that in the days when our grandfathers worked steel I made heading dies of two inches diameter by upsetting 1 ¼-inch octagon steel, which stood to do as much work as any that I could make from steel of other shape and size of hundreds which were made in this way, I do not remember that seams or cracks resulted from upsetting in a single instance. If the steel was sound, and the work in upsetting was so distributed as to affect all parts of the mass alike, why should they?

As to upsetting cold chisels, strips of sheet-steel, iron, or pieces of steel of any shape which are so slender as to double up like straws from the power of the blow, there are not enough blacksmiths wasting time and injuring steel in that way to call for any protest. Steel is hammered with a view to improve it many times, in these latter days, in a manner so much like that in vogue with the "old chaps" that it is a wonder that so many of them have ceased from labor without ever having been told that they had "monkeyed" steel. Well, if the devil has only got those smiths who hammered cold steel, there are not many smiths in his ranks yet. Heating steel as hot as it will bear and working it while hot, with injunctions not to try to draw cold steel, never to upset, may be construed to mean: hot as is best for steel, no hotter; not cold, but continue hammering till the grain of the steel is well closed up and finer for hammering at a low heat than it could possibly be made if the hammering was all done while the steel was hotter. It don't seem that everybody does write in the papers against heating steel too hot, and nobody can come to the conclusion that half the trouble is caused by too little heat, if his experience and opportunities for observation have been the same as mine.

What a difference it makes where a chap, old or young, stands to look at a thing. But I am off the track. Upsetting steel is the question. The screw blanks which were headed in the upset dies were threaded with rotary cutters. These cutters may be described as a section of a worm gear, the shape of tooth being that of the space between threads on a wood screw, and the cutter having a rotary motion in keeping with the rotary motion of the screw and with its own longitudinal motion as it traversed the length of the screw which it cut. The best of steel, according to our best judgment, was used for these cutters, but running through many of the best brands, from "Jessup's" to "Hobson's Choice Extra Best," these cutters did not stand as it seemed they ought to do. I was anxious to try a high grade of a certain brand, of which I could get none of the right size till it could be imported (this was in the days of the old chaps—grandfathers). I had some one-half of an inch or five-eighths of an inch round; the cutters were to finish round, over three-fourths of an inch diameter. I could not wait, so I made some cutters by upsetting the steel which I had. These cutters stood to do five times the work of the cutters I had been using; and, willing to "let well enough alone," I continued using the same size of steel and upsetting.

I once had for my job the lathe work on the hook for the Wheeler & Wilson sewing machine. As it came to me it was nothing more nor less than a round-headed steel bolt, made in just the same way that a blacksmith would make any solid headbolt, by drawing a body to pass through the heading tool and upsetting the head. After this lapse of time— thirty years— I can hardly be expected to remember the character of each particular hook, but I do not remember that one of them was unsound as a result of upsetting. If steel gets too much work in one part and not enough in another, in any direction, hot or cold, it is likely to get broken up. Upsetting need not take all the blame.

How much difference in fracture does anyone suppose will be found between breaking a bar of steel crosswise and breaking a piece of the same length of the width of the bar the other way of the grain? Cut two pieces of steel, each two inches long, from a two-inch square bar and draw one of them down to one inch square, or eight inches long in the direction of the length; draw the other piece to the same size and length by upsetting at every other blow, or drawing crosswise. Who believes that the tool made from this last piece will be inferior to the one made the right way of the grain? Some bars of steel, and steel for some uses, can be improved by the work which they can get on a smith's anvil. Some bars of steel, and steel for some uses, will be injured by the same treatment. When it is necessary or desirable to use the center of a bar of steel, and it is found to be in a coarse condition by reason of having been brought to shape and size by blows disproportioned to its cross-section, there is, perhaps, no better way of improving the grain at and near the center of the bar than to upset, and alternate the upsetting with work from the outside of the bar.

Upsetting is not a cheap, easy or desirable thing to do, and is not likely to be resorted to often except when necessity compels; but I see no use in making a bugbear out of it, when, if properly done, there is no harm in it, and in some cases actual good is done by it. It does not at least break up the center of the bar, as blows at right angles with its axis often do; and, after seeing, as I have, thousands of good cutters made by upsetting, for uses requiring the utmost soundness and best condition of steel, I do not like to see the process tabooed without saying a word in its favor. —S. W. Goodyear, in *The American Machinist*.

HEATING STEEL IN THE BLACKSMITH'S FIRE.

In heating steel, two faults are especially to be guarded against. First, overheating; secondly, unequal heating of the parts to be operated on (whether by forging or tempering). Referring to the first, many blacksmiths do not recognize that steel is burned unless it falls to pieces under the hammer blows, whereas that condition is only an advanced stage of the condition designated as *burnt*. This is the secret of their partial failure, or, that is to say, of the inferiority of their work. Others recognize that from the time a piece of steel is *overheated*, to its arrival at the stage commonly recognized as *burnt*, a constant deterioration is taking place.

In the practice of some, this excessive heating may be carried to so small a degree as not to be discernible, except the tool be placed in the hands of an operator whose superior knowledge or skill enables him to put it to a maximum of duty, less skillful manipulators being satisfied with a less amount of duty.

But in the case of stone-cutting tools especially, and of iron cutting tools when placed in the hands of very rapid and expert workmen, the least overheating of the tool will diminish its cutting value as well as its endurance. A piece of steel that is burnt sufficiently to break under the forging hammer, or to be as weak as cast iron, will show, on fracture, a coarse, sparkling, granulated structure, and this is the test by which working mechanics, generally, judge whether steel is burned or not. But it is totally inadequate as a test to determine whether the steel has not suffered to some extent from overheating. Indeed, although the grain becomes granulated and coarse in proportion as it is overheated, yet it may be so little overheated as to make no *visible difference in the grain of the fracture, although very plainly perceptible in the working of the tool, if placed in the hands of a thoroughly good workman.*

When the results obtained are inferior, it is usual to place the blame on the steel, but in the case of well-known brands of steel, the fault lies, in ninety-nine cases in a hundred, in over-heating, either for the forging or for the tempering.

In determining from the duty required of it, whether a tool comes up to the highest standard of excellence, the best practice must be taken as that standard. Thus, if it is a metal cutting tool, as, say, a lathe tool, let the depth of cut be that which will reduce, at one cut, say, a four-inch wrought-iron shaft down to 3 ¼ inch diameter, the lathe making, say, 16 revolutions, while the tool travels an inch, and making from 25 to 30 revolutions per minute. Under these conditions, which are vastly in excess of the duty usually assigned in books to lathe tools, the tool should carry the cut at least four feet along the shaft without requiring grinding.

If the tool is for stone work, let it be tested by the most expert and expeditious workman. These instructions are necessary because of the great difference in the quantity of the work turned out in the usual way and that turned out by very expert workmen.

At what particular degree of temperature steel begins to suffer from overheating cannot be defined, because it varies with the quality of the steel. The proper degree of heat sufficient to render the steel soft enough to forge properly and not deteriorate in the fire is usually given as a *cherry red*, but this is entirely too vague for entirely successful manipulation, and, in practice, covers a wide range of temperature. The formation of scale is a much better test, for when the scales form and fall off of themselves, the steel, in fine grades of cast steel, is overheated, and has suffered to some extent, though the common grades of spring or machine steel may permit sufficient heat to have the scale fall off without the steel being worked. As a rule, the heat for tempering should be less than that for forging, and should not exceed a blood red. There are special kinds of steel, however, as, for example, chrome steel, which require peculiar heating, and in using them strict attention should be paid to all instructions given by the manufacturers.

Steel should be heated for forging as quickly as compatible with securing an even degree of heat all through the part to be forged, and heated as little as possible elsewhere. If this is not done the edges or thin parts become heated first, and the forging blows unduly stretch the hottest parts, while the cooler parts refuse to compress; hence a sort of tearing action takes place, instead of the metal moving or stretching uniformly.

The steel should be turned over and over in the fire, and taken frequently from the fire, not only to guard against overheating, but because it will cool the edges and tend to keep the heat uniform.

The fire may be given a full blast until the steel begins to assume redness at the edges or in the thin parts, when the blast must be reduced. If the thin part is heating too rapidly it may be pushed through the fire into the cooler coals or taken out and cooled in the air or in water, but this latter should be avoided as much as possible.

When the steel is properly heated, it should be forged as quickly as possible. Every second of time is of the utmost importance.

There must be no hesitation or examining while the steel is red-hot, nor should it be hammered after it has lost its redness. There is, it is true, a common impression that by lightly hammering steel while black-hot it becomes impacted, but this only serves to make the steel more brittle, without increasing its hardness when hardened.

If the tool has a narrow edge, as in the case of a mill pick or a chisel, the first hammer blows should be given on those edges, forging them down at first narrower than required, because forging the flat sides will spread the edges out again. These edges should never be forged at a low heat; indeed, not at the lowest degree of red heat, or the steel at the outer edge is liable to become partly crushed.

What is known as jumping or upsetting—that is, forging the steel endways of the grain should be avoided, because it damages the steel.

As the steel loses its temperature, the blows should be delivered lighter, especially upon the edges of the steel.

The hammer blows for drawing out should have a slight lateral motion in the direction in which the steel is to be drawn, so that the hammer face, while meeting the work surface, fair and level, shall also draw the metal in the lateral direction in which the face of the hammer is moving at the moment of impact.—*By* Joshua Rose, M. E.

TESTING IRON AND STEEL.

The English admiralty and "Lloyds'" surveyor's tests for iron and steel are as follows:

Two strips are to be taken from each thickness of plate used for the internal parts of a boiler. One-half of these strips are to be bent cold over a bar, the diameter of which is equal to twice the thickness of the plate.

The other half of the strips are to be heated to a cherry red and cooled in water, and, when cold, bent over a bar with a diameter equal to three times the thickness of the plate—the angle to which they bend without fracture to be noted by the surveyor. Lloyds' circular on steel tests states that strips cut from the plate or beam are to be heated to a low cherry red, and cooled in water at 82° Fah. The pieces thus treated must stand bending double to a curve equal to not more than three times the thickness of the plate tested. This is pretty severe treatment, and a plate containing a high enough percentage of carbon to cause any tempering is very unlikely to successfully stand the ordeal. Lloyds' test is a copy of the Admiralty test, and in the Admiralty circular it is stated that the strips are to be one and a half inches wide, cut in a planing machine with the sharp edges taken off.

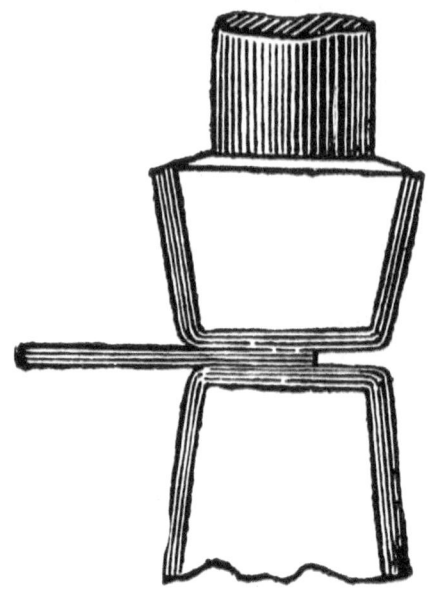

FIG. 3—CLAMPING IN A STEAM HAMMER
FOR THE PURPOSE OF BENDING.

One and a half inches will generally be found a convenient width for the samples, and the length may be from six to ten inches, according to the thickness of the plate. If possible, the strips, and indeed all specimens for any kind of experimenting, should be planed from the plates, instead of being sheared or punched off. When, however, it is necessary to shear or punch, the pieces should be cut large and dressed down to the desired size, so as to remove the injured edges. Strips with rounded edges will bend further without breaking than similar strips with sharp edges, the round edges preventing the appearance of the small initial cracks which generally exhibit themselves when bars with sharp edges are bent cold through any considerable angle. In a homogeneous material like steel these initial cracks are very apt to extend and cause sudden fracture, hence the advantage of slightly rounding the corners of bending specimens.

In heating the sample for tempering it is better to use a plate or bar furnace than a smith's fire, and care should be taken to prevent unequal heating or burning. Any number of pieces may be placed together in a suitable furnace, and when at a proper heat plunged into a vessel containing water at the required temperature. When quite cold the specimens may be bent at the steam-hammer, or otherwise, and the results noted.

The operation of bending may be performed in many different ways; perhaps the best plan, in the absence of any special apparatus for the purpose, is to employ the ordinary smithy steam-hammer. About half the length of the specimen is placed upon the anvil, and the hammer-head pressed firmly down upon it, as in Fig. 3. The exposed half may then be bent down by repeated blows from a fore-hammer, and if this is done with an ordinary amount of care it is quite possible to avoid producing a sharp corner. An improvement upon this is to place a cress on the anvil, as shown at Fig. 4.

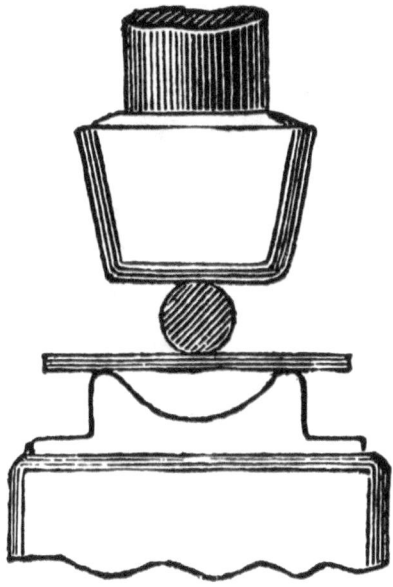

FIG. 4. —BENDING UNDER A STEAM HAMMER OVER A HOLLOW ANVIL AND BY MEANS OF A ROUND BAR.

The sample is laid upon the cress, and a round bar, of a diameter to produce the required curve, is pressed down upon it by the hammer-head. The further bending of the pieces thus treated is accomplished by placing them endwise upon the anvil-block, as shown in Fig. 5. If the hammer is heavy enough to do it, the samples should be closed down by simple pressure, without any striking. Fig. 6 is a sketch of a simple contrivance, by means of which a common punching machine may be converted temporarily into an efficient test-bending apparatus. The punch and bolster are removed, and the stepped cast-iron block, *A*, fixed in place of the bolster. When a sample is placed endwise upon one of the lower steps of the block *A*, the descending stroke of the machine will bend the specimen sufficiently to allow of its being advanced to the next higher step, while the machine is at the top of its stroke. The next descent will effect still further bending, and so on till the desired curvature is attained. It would seem an easy matter, and well worth attention, to design some form of machine specially for making bending experiments; but with the exception of a small hydraulic machine, the use of which has, I believe, been abandoned on account of its slowness, nothing of the kind has come under the writer's notice.

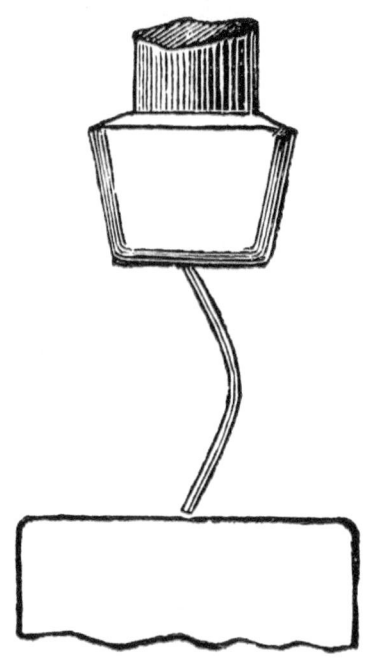

FIG. 5—BENDING STILL FURTHER BY MEANS OF A STEAM HAMMER.

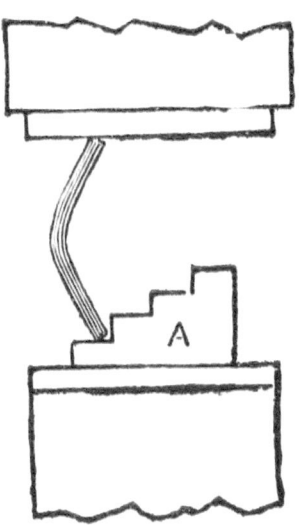

FIG. 6—A SIMPLE CONTRIVANCE BY WHICH A COMMON PUNCHING MACHINE MAY BE CONVERTED INTO A TESTING APPARATUS.

The shape of a sample after it has been bent to pass Lloyds' or the Admiralty test is shown at Fig. 7. While being bent the external surface becomes greatly elongated, especially at and about the point A, where the extension is as much even as fifty percent.

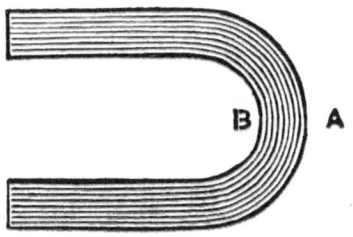

FIG. 7—SHAPE OF A SAMPLE AFTER BEING BENT TO PASS THE ADMIRALTY TEST.

This extreme elongation corresponds to the breaking elongation of tensile sample, and can only take place with a very ductile material. While the stretching is going on at the external surface, the interior surface at B is being compressed, and the two strains extend into pieces till they meet in a neutral line, which will be nearer B than A with a soft specimen. When a sample breaks the difference between the portions of the fracture which have been subject to tensile and compressive strains can easily be seen. Fig. 8 shows a piece of plate folded close together; and this can generally be done with mild steel plates, when the thickness does not exceed half an inch.

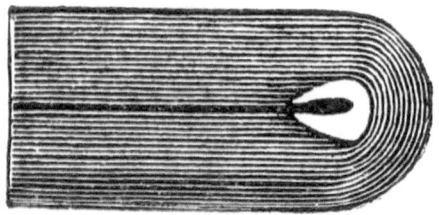

FIG. 8—A PIECE OF PLATE FOLDED CLOSE TOGETHER.

Common iron plate will not, of course, stand anything like the foregoing treatment. Lloyds' test for iron mast-slates ½ inch thick, requires the plates to bend cold through an angle of 30° with the grain, and 8°. across the grain; the plates to be bent over a slab, the corner of which should be rounded with radius of ½ inch.

THE TREATMENT OF STEELS.

I think it possible that some ideas of mine on the subject of the treatment of steel may be of interest, but I do not make these remarks for the purpose of attacking any opinions that have been advanced on this matter by others.

My purpose is simply to call attention to some important facts which appear to have been unknown or overlooked by the advocates of certain methods of treating steel.

First, let me say that I have no faith in the idea that the operator can be guided correctly by color in the tempering or hardening of steel. If the word steels is used instead of steel, and if the teacher in the mysteries of hardening and tempering is thoroughly familiar with the character of each particular steel—and their name is legion, with the exact heat at which each would harden to the best advantage, and the precise color at which the temper should be drawn—this would not be enough to justify him in laying down the color or any other specific rule. He should also be able to make millions of pupils see a color exactly alike, use the same kind of fuel, work in the same light and remember the precise heat and color necessary for each particular steel. Then, and not till then, he may speak of specific rules.

Is it asked what can be done if no specific method will apply? The reply is, that with so great a variety of steels and such a variety of grades of the same steels, with such a variety of treatments required for pieces of different sizes of precisely the same steel, we must look for successful hardening and tempering to the intelligence and powers of observation and comparison to be found among those who do this most important part of tool-making.

These rules will, however, apply in every case. Try to learn the lowest heat at which each particular steel will harden sufficiently to do the work for which it is intended, and never exceed that heat. Harden at a heat which calls for no drawing down to a blue to remove brittleness. Hardened at the proper heat, steel is stronger without drawing the temper than it is before hardening. Whenever hardened steel snaps off like glass at a little tap, and shows a grain as coarse, or coarser than it showed before hardening, don't say "It is too hard," but say "It was too hot." Steel is hardened by heating to the proper heat and cooling suddenly. Good results can never be obtained by heating steel too hot for sudden cooling, and then cooling in some mixture or compound which will cool the overheated steel more gradually. Overheating steel does mischief which cooling gradually only partially removes. Many a man who hardens in some mixture or compound to prevent cracks or distortion, may learn that at a lower heat he may use a bath of cold water with equal safety and better results. —*By* S. W. G.

HINTS ABOUT THE TREATMENT OF STEEL.

A practical worker in steel gives the following hints in reference to the treatment of tool steel.

He says: "Bosses of machine and other shops where considerable steel is used would be astonished could they see the amount of loss to them in broken taps, dies, reamers, drills and other tools accumulated in one year. I do not mean to say that tools can be made and hardened that will not break, but I do say fully one-third, especially taps, are broken by being improperly tempered. Thousands of dollars are lost in waste of steel and loss of time by men who try to do what they have never learned".

"First of all, never heat steel in any fire, except a charcoal fire, or a lead bath, which gives it a uniform heat (I mean steel tools of any kind). Rainwater is the best water for hardening, as it contains less lime than either well or river water. Water should never be very cold for hardening tools, as the sudden contraction of the steel will cause it to crack, more especially in cutting dies, where there are sharp corners."

"To insure thorough hardness, salt should be added to the water until the same becomes quite brackish. The salt will cause the water to take hold of the steel and cool it off gradually. I find rock or fish salt to be the best, though table salt will answer. The salt is also beneficial to the carbon in steel."

"For removing scale from steel when put into water, ivory black should be used, putting it on the steel while it is heating, and letting it remain till the steel goes into the water."

"In tempering taps, reamers, twist drills and other like tools, great care should be taken to put them in the water in a perpendicular line, and slowly, not allowing them to remain stationary in the water, as there is danger of there being a water crack at the water line."

"A good way to draw the temper is to heat a collar, or any suitable iron with a hole in the center, and draw the tool backward and forward through it until the right temper is obtained, which will be uniform."

ON THE WORKING OF STEEL.

Allow me to express a few thoughts in regard to the degree of heat steel can be worked for edge tools. And in doing so I shall no doubt differ with many; but what of that? If we all had one way of doing work what would be the use of giving "*our method*" for doing anything? Some tell us that steel should never be heated above a cherry-red, others say it should not be hot enough to scale, and many suppose if heated to a white heat the steel is burnt and utterly worthless. Now if all this be true, how in the world could an edge tool ever be made, and of what practical use would they be when they *were* made?

Can steel be put into an axe or any other tool without heating the steel above a cherry-red heat? If so, I would like to have some one tell me how, as I have never learned that part of the trade. I do not believe that iron or steel can be welded at a cherry heat.

My experience in the working of steel during the past fifteen years has been mostly confined to the axe business. I have made and repaired during this time, several hundreds of axes, and the work has given unparalleled satisfaction, My experience in the matter has convinced me that the degree of heat steel should be worked depends very much upon circumstances. For instance, if I am going to fix over an axe, and wish to reduce the steel to one-half or three-fourths of its present thickness, I have no fears of any bad results if the steel is brought to a white heat to commence with. But when nearly to the required thickness, I am careful not to heat above a cherry-red. And when the last or finishing touch is given by the hammer it is at a low heat, when but a faint red is discernible. I never finish forging an edge tool of any kind at a cherry-red heat. The finishing should be done at as low a heat as to refine the steel, and leave it bright and glossy.

In heating to temper, the greatest care should be observed that an even cherry wood heat is obtained. I do not deny that injury may occur by overheating. This every smith knows to be true; but I *do* claim that it can be remedied, and the fine grain of the steel restored when the nature of the work will admit of a suitable amount of forging. When it will not, I never heat above a cherry-red. —*By* W. H. B.

HARDENING STEEL.

On the subject of the hardening of steel I will say that salt water is not more liable to crack steel in the hardening process than is fresh water. In fact, clear and pure water is the best thing that can be used in making steel hard.

The best mode of annealing heavy blocks of crucible cast-steel is to heat the block to a uniform red heat, and as soon as you have obtained this heat place the steel in a cast-iron or sheet-iron box, made as shown in Fig. 9, of the accompanying illustrations. This box is filled with common lime and wood ashes, equal parts of each. In placing the steel in the box try to keep about 4 or 5 inches of lime below the steel, then put 4 or 5 inches on top of the steel, close up the box and let it remain there to cool off slowly, which will require from 24 to 48 hours. Generally crucible cast-steel will anneal best at a low red heat, if treated as described. The only secrets in annealing steel, that I know of, are to exclude air from the steel as much as possible, while it is annealing, and to avoid overheating.

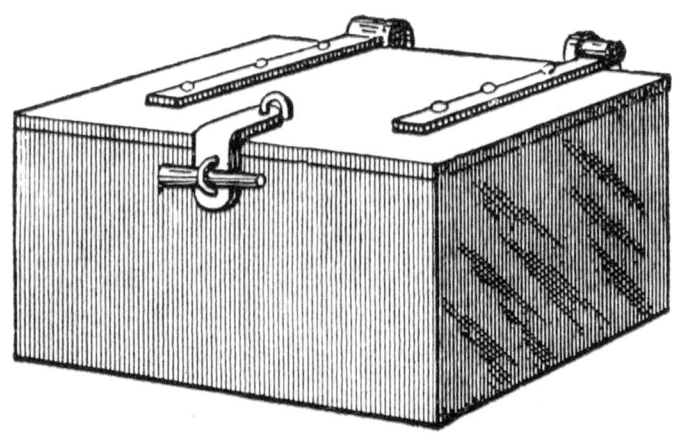

FIG. 9—SHOWING THE BOX USED IN HARDENING STEEL BY THE METHOD OF "H. R. H."

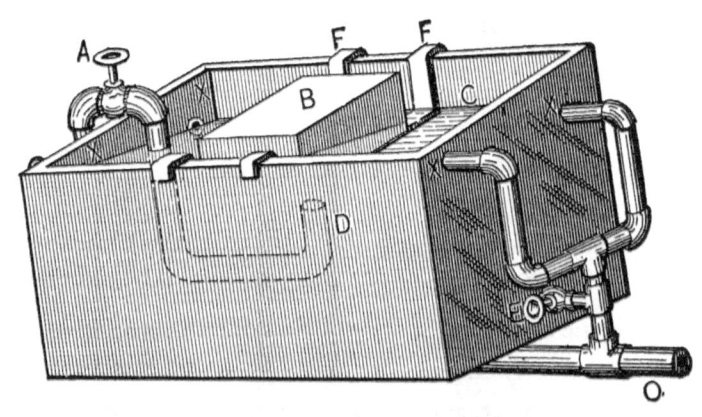

FIG. 10—SHOWING THE WATER TUB.

For hardening drop hammer dies, I would suggest the use of the water tub, shown in Fig. 10. In the illustration, *A* is the water supply pipe, which, as shown by the dotted lines, runs down to the bottom of the tub and over to its center, where the pipe is bent up, as shown at *D*. At *x, x, x, x* there are four holes, which represent waste pipes that carry off all the water that comes above their level. This water passes on down through to the main waste pipe *O* at *E*. The cut-off shown is used when cleaning the tub or box.

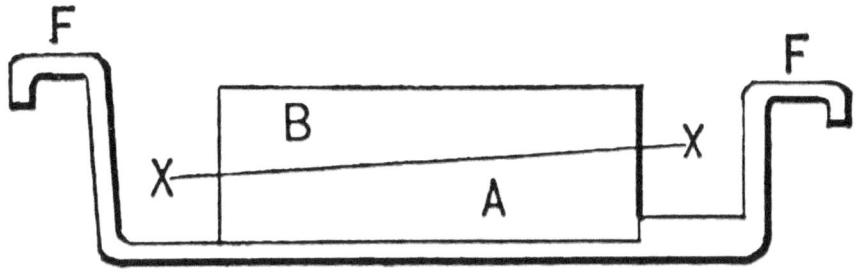

FIG. 11—SIDE VIEW OF THE DIE.

To harden a heavy die proceed as follows: Get a uniform bright red heat on the steel, then place it in the water tub with the face of the die in the water, the steel being supported in the tub by the two hangers, *F F*, as shown in Fig. 10. *C* indicates the water's edge and depth, and shows the position of the die in the water. Before placing the die on the hangers be sure to have the two hangers in their proper places, have the water in the supply pipe, *A*, turned on with full force, and be sure that the die is placed directly over the middle of the supply pipe at *D*. We will say the die *B*, in Fig. 10, is 10 x 5 x 6 inches; a bulk of steel like this will heat much water, and therefore while the supply pipe furnishes cold water, the waste pipes, *x, x, x, x*, carry off all the hot or warm water as fast as the steel heats it. In Fig. 11 I give a side view of the die on the hanger, *F, F,* showing the ends that hook over the sides of the tub.

The cut also shows how the die is placed at an angle in the water. *B* indicates the portion of the die out of the water line, and A represents the part in the water. It is not necessary to keep the dies moving during the hardening process, but keep the water moving, and there will be no trouble.

The best way to harden trimming dies is to always examine thoroughly the die before you dip it, and turn it while dipping so that all the thick or heavy parts enter the water first.

FIG. 12—THE FURNACE FOR HEATING AND DRAWING TEMPER.

The best mode of hardening punches is to harden them first in cold water, and then draw the temper to suit the work they are intended for. While the die is hanging in the tub, as in Fig. 10, it would do no harm to utilize a tin. cup in pouring water on the bottom of the die, which is out of the water. I have seen hundreds of blocks of steel hardened this way with the best of success. I never have the temper drawn for drop hammer dies. The harder you can get them the better.

In Fig. 12 I show a furnace for heating and drawing temper. This furnace is 3 feet 6 inches wide and 4 feet 4 inches long. At E it is 2 feet 6 inches from the ground to the top of the cast-iron plate B. From plate B to the underside of the top, F, it is 16 inches in height.

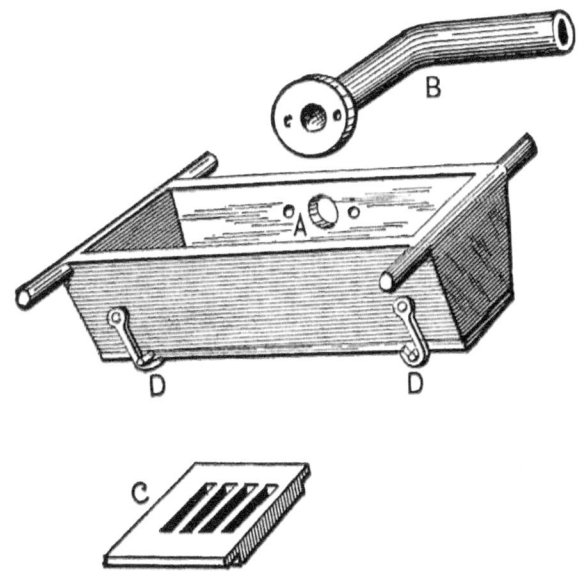

FIG. 13—SHOWING THE GRATE-BOX, BLAST-PIPE AND GRATE.

A is the grate, D is the ash-pan, C is the door, and X is the damper. Hard coke is the best fuel I know of for this furnace except charcoal. In using the coke start the fire, close the door and get the coke red then put in the steel that is to be annealed or hardened. In heating you can gain much by regulating the damper X. Fig. 13 shows the grate box used in the furnace. At D, D, the box has two hooks, which hold up the box door. This door is fastened to the other side of the box with two hinges. At A is a blast pipe hole to which the point of pipe B is bolted. C is part of the grate used. Three pieces of this kind are used to make the full grate. This is done to allow the use of two blank grates and one open grate when a small fire is desired, or you can use two open grates and one blank, or all three may be open grates, just as your work requires. Fig. 14 shows the plate in which the grate box is hung when in use. This plate is bolted to two iron bars in the furnace, and on this place the brick work is laid. The chimney can be used on the side, end or middle. I prefer it in the center, as shown in the illustration.

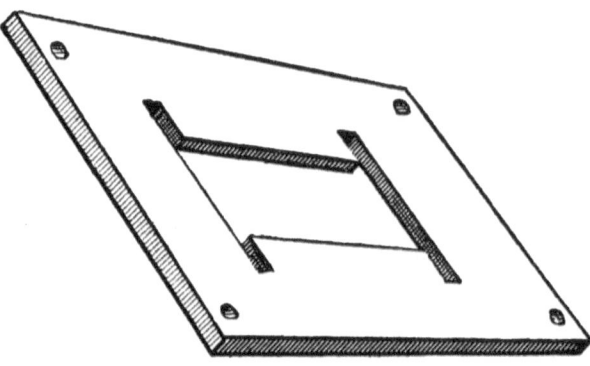

FIG. 14—THE PLATE FOR HOLDING THE GRATE-BOX.

The plate, *B*, of the furnace is cast-iron 1 ¼ inches thick, 10 inches wide, and the length is the same as the width of the furnace. The best steel I can get for drop and trimming dies is made at the Crescent Steel Works, Pittsburgh, Pa. I think the illustrations fully explain themselves without any further description—*By* H. R. H.

TO SELECT GOOD TOOL STEEL.

One way is to break a bar of steel and observe the grain, which should be fine and present a silvery look, with sometimes an exfoliated appearance. The best test of steel is to make a cold chisel from the bar to be tested, and, when carefully tempered, try it upon wrought iron, a piece of old wagon-tire, for instance. The blows given will pretty correctly tell its tenacity and capability of holding temper. If it proves tough and serviceable take this temper as a guide, and temper your steel in like manner. Inferior steel is easily broken, and the fracture presents a dull, even appearance, which might be appropriately termed a lifeless look. —*By* W. B. H.

DIFFERENT KINDS OF STEEL.

Blister steel is made by causing the carbon of charcoal to penetrate iron in a heated state. German steel is blister steel rolled down into bars. Sheet steel is made by hammering blister steel. Double shear steel is made by cutting up blister steel and putting it together and hammering again. Crucible steel is made by melting in a pot blister steel and wrought iron or unwrought iron and charcoal and scrap. Bessemer steel is made by blowing air through cast-iron, burning out the silicon and carbon. Open hearth steel is made by melting pig-iron and mixing wrought iron or scrap steel, or iron ore to reduce the silicon and carbon.

RESTORING BURNT STEEL.

To restore the original qualities to steel which has been burnt in the forge, plunge the metal at red heat into a mixture of two parts of pitch, two parts of train oil, one part of tallow, and a small quantity of common salt. Repeat the operation two or three times. Excellent results have frequently been so obtained.

COLD HAMMERING IRON.

To the statement by a writer on this subject that "it either is, or ought to be known to all practical men that hammering a piece of the best and toughest iron in the process of forging until it ceases to be red hot, will remove and destroy its tenacity so as to render it capable of being broken with the slightest blow," practical men must say, "depends."

It depends upon the character of the iron, and upon how the hammering is done. As between hot working, and the finishing blows "of cold hammering," i. e., hammering at black heat—not cold—there are two reasons why the effect so strongly deprecated is produced.

First, the iron is less yielding in this semi-cold state and so would not be affected clear through, or as nearly through, by the same blows it received when hot.

Second, the blows during the cold hammering are light compared with what were used when the heat was greater. Thus, if the best condition of the mass is considered, we have the heaviest blows when least force is needed, and the lightest blows when to move the mass the heaviest are needed, and so, while the "requisite finish and fine surface" result from the cold hammering there is an evil effect produced, not from the hammering *per se*, but, from the fact that only the surface being affected to any considerable extent, the desirable homogeneity of condition is destroyed and unequal strains are set up which can only be relieved by annealing.

Quoting again, "By subjecting wrought-iron to the most violent hammering or compression at a low temperature, and then submitting the iron work so treated to the simple process of heating red hot and slow cooling, we enhance its tenacity, or shock sustaining qualities at least twenty times."

Now, without questioning the accuracy of this statement, is it not fair to ask if cold working is done in a way to affect the entire mass acted upon clear through, putting all parts as nearly as possible in the same condition would there not be *greater* "tenacity or shock-sustaining qualities" *without* subsequent annealing?

In cold working of both wrought iron and steel, the writer has had to do with, and opportunity for observing the effect of reduction as great as from 25 per cent. to 75 per cent. often from drawn wire, not annealed after drawing, and without heating or annealing after such cold working, millions of these pieces have been bent, flattened, riveted and otherwise treated in a way to test their tenacity, without showing any sign of having had the "tenacity removed or destroyed," but on the contrary greatly increased, while actual tests for tensile, torsional, or transverse strength showed great increase in these directions, but which increase would, in a great measure, be lost by "heating red hot and slow cooling." If, in the article under consideration, the term cold-hammering had been used only, this would not have been written, but, as the terms "swaging" and "compression" were used, the door was opened. What is the difference? It is immense in its effect, as between simple hammering and swaging, between compression—squeezing—blows and hammering.

Hammering implies working between two plain faces which allows some parts of the metal acted upon to escape from the compressive effect of the blows more easily than other parts, hence unequal conditions result. Swaging implies the use of dies, which hold all parts of the metal acted upon up to the work they are to receive, and so produce an equable condition all through the mass. Again, compression, as against blows, produces effects peculiar to itself in that the work takes place in a gradual, gentle manner, rather than through shock and violence. Just why there is so marked a difference, whether it is because the parts composing the mass having more time are able to arrange themselves differently from what they do under the sudden effect of blows, whether the less friction of changing parts and less consequent heat, or any other of many guessed at causes lie back of what strikes the average mechanic as a phenomena, a paradox, will probably remain an open question for some time to come. That compressive swaging, properly done, however, will increase tenacity and strength tested in any way we choose, by bending, twisting, pulling, etc., is an unquestionable fact.

Pieces of cast-steel wire of high carbon percentage, suitable for drills, have been reduced by cold swaging sufficiently to become elongated more than 700 per cent, and then tied in knots and drawn up almost as tightly as would be possible in the case of a string. That this could be done when the fact is taken into consideration that the wire had been cold-worked—drawn since annealing—and was consequently in that condition so deprecated in the article under notice, the subsequent cold-working—swaging —taking place without annealing or heating after the drawing, and the knots being tied after the swaging with no heating or annealing, should settle the question, to some extent at least, whether cold-working per se, is the destructive agent which some believe it to be. —*By* S. W. Goodyear.

CHAPTER II.

BOLT AND RIVET CLIPPERS.

A BOLT AND RIVET CLIPPER.

Cutting off bolts and rivets with a cold chisel is not very convenient in a shop where only one man is working: for instance, a blacksmith shop in a small town. Very good bolt and rivet clippers are now manufactured, but many blacksmiths cannot afford to pay eight or ten dollars for a bolt clipper, and so they have some one to hold a hammer or bar on one side of the bolt while the smith cuts from the other side with a dull chisel, and now and then hits his hand, or the end of the bolt flies in his eye or in the eye of the man that holds the bar.

FIG. 14—SHOWING HOW THE KNIFE IS MADE FOR BOLT AND RIVET CLIPPER.

Then very often the end of the bolt goes through a window, and before they get through with their job the smith is very mad.

About three years ago I made a good and cheap bolt clipper, which is shown in the accompanying illustrations, Figs. 14 to 17. It is made as follows: A piece of steel ½ x 1 inch and 6 inches long is welded to a ¾-inch round rod 12 inches long, and the end of the steel is turned up half an inch for a nipper or knife, as at *A* in Fig. 14.

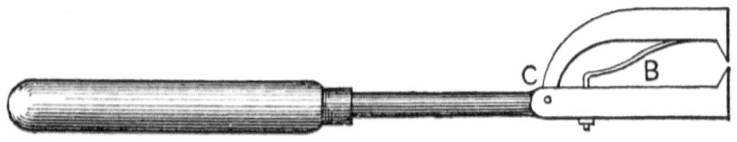

FIG. 15—SHOWING THE TWO JAWS TOGETHER.

In Fig. 15 the two jaws or nippers are together. *B*, in Fig. 15, is a spring used to raise one jaw when the tool is applied to a rivet. The upper jaw works loosely in a slot hole at *C*. A small hole is punched six inches from the end for the spring. A nut is used to fasten the spring. In Fig. 16 the purchase lever is shown. This is made of inch-square iron and 12 inches of ¾-round iron, or just as the lower handle is made.

FIG. 16—THE PURCHASE LEVER.

A ½-inch round hole is punched in one side at *D*, as in Fig. 16. In Fig. 17 the clipper is shown as it appears when put together and ready to be applied to a bolt.

Fig. 17 is a side view of the clipper. The jaws, *F, F*, must be close to the piece *W*. When you press down the lever the lower side of the head commences to press down. This clipper can be made in half a day and will answer for most jobs. —*By* E. H. Wehry.

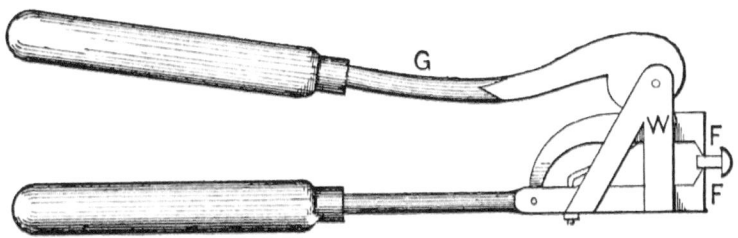

FIG. 17—SHOWING THE CLIPPER COMPLETED.

CUT NIPPERS.

I send you a sketch, Fig. 18, of a pair of cut nippers I invented. They are not patented, nor will they be. Three-eighths-inch iron can be cut with them with ease.

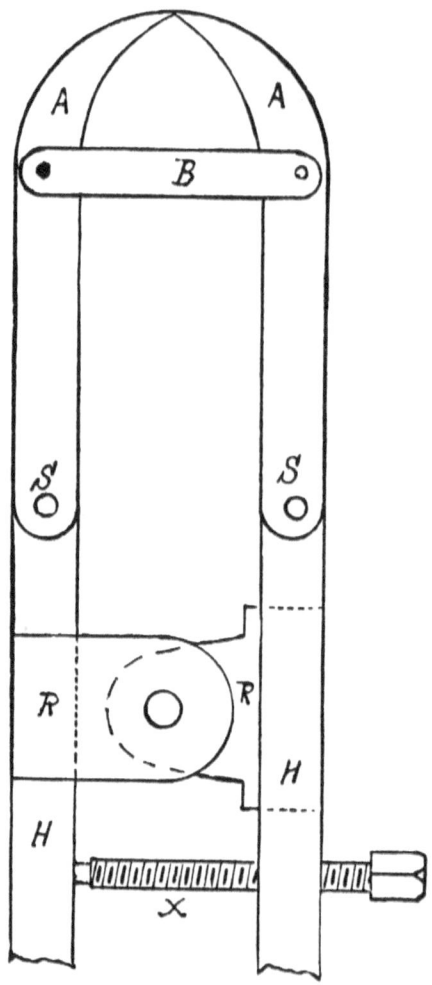

FIG. 18—CUT NIPPERS AS MADE BY "STEEL SQUARE".

A A are steel cutters down to joints *S S*, and they may be made of any shape to suit. At *B* there are two links bolted to the cutter, one on each side. The joint *R R* would be difficult to forge, so it is made of malleable iron, and is bolted on the side of the handle *H*. There are two of these made with a shoulder on the inside. The right-hand part is bolted on the edge of the other handle *H*, is the same thickness as the handle and sets in between the other two, being held by a bolt. A set screw, as shown, stops the handles *H H* at the right point. The handles may be of any length desired. —*By* Steel Square.

BOLT CLIPPER.

I inclose sketches illustrating a bolt clipper which may be made by any good blacksmith in four hours' time. Fig. 19 represents the tool complete, while the other sketches represent details of construction. For the parts shown in Fig. 21 take a piece of spring steel 2 ¼ inches wide by ¼ inch thick and flatten out about 2 ½ inches wide at *B*.

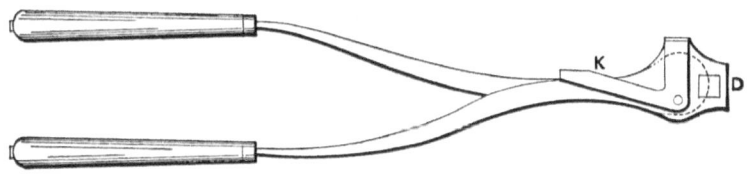

FIG. 19—THE BOLT CLIPPER COMPLETE.

That will leave the part 3-16 of an inch thick. Punch holes as shown at *A* and *B*. Shape a small piece of steel as indicated by *C* in the same cut and place it on the end.

FIG. 20—PORTION OF BOLT CLIPPER. ELEVATION, SECTIONAL VIEW AND DETAILS OF "D. H. E.'S" BOLT CLIPPER.

FIG. 21—ANOTHER PART OF BOLT CLIPPER.

Take a light heat and weld it fast in that position. That will keep the end from pushing out. The square hole marked *O* in Fig. 21 is made large enough to pass it over the nuts. The part shown in Fig. 20 is made of cast-steel and sharpened in the parts shaded as shown at *H*. The construction of the guard is shown in Fig. 22. It is to be bent at the dotted lines, giving it the shape indicated by Fig. 23. It is then ready to clinch into the holes provided for it as shown in *A* and *B* of Fig. 21.

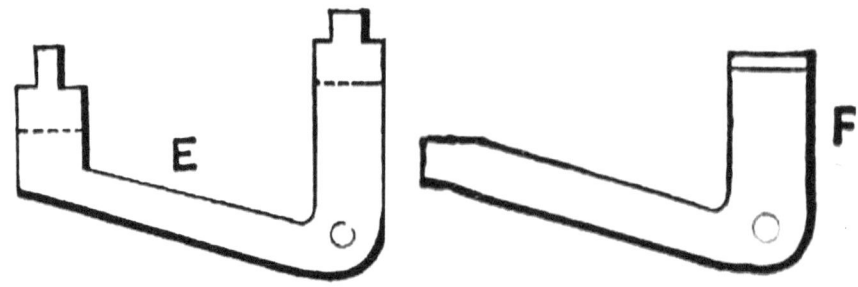

FIG. 22—SHOWS HOW THE GUARD IS CONSTRUCTED

FIG. 23—SHOWS SHAPE OF GUARD

The bolt uniting the two parts should be made of cast-steel 5-16 inch in diameter. The entire length of the tool should be 15 inches. Made of these dimensions leverage enough will be afforded to clip bolts 3-16 to 5-16 inches in diameter. —*By* D. H. E.

A NEW BOLT CLIPPER.

I enclose a sketch, Fig. 24, of a bolt clipper which is a handy tool and unlike any I have seen in other shops. The handles are of wood, and are about two feet long.

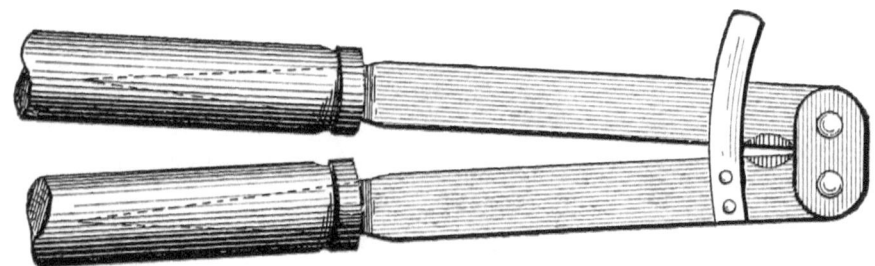

FIG. 24—A NEW BOLT CLIPPER AS MADE BY "R. D. C."

The band or clamp prevents the twisting of the knives to one side when they close on the bolt. The plate shown in the sketch is duplicated on the other side. This arrangement enables me to get a leverage near the hinge or heel. This tool can be used for bolts ranging in size from the smallest up to half inch. — *By* R. D. C.

A HANDY BOLT CUTTER.

I enclose sketches of a bolt cutter of my own make, which I will describe as well as I can. I think the tool may be of some benefit to some of my brother smiths.

It saves labor and is easily made. To make it I first take a bar of iron 7-8-inch square and cut off two pieces, each two feet long, for the levers *A* and *B* shown in the engravings, Figs. 25 and 26.

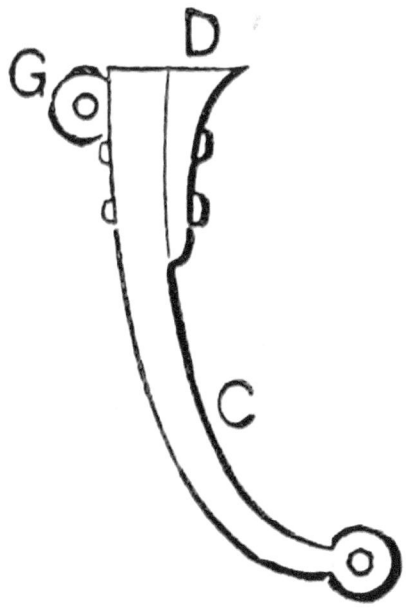

FIG. 25—A HANDY BOLT CUTTER. THE BOW.

In making the lever A, I first square up the end where the hole, *G*, is made. I then punch, six inches below the hole *G*, another hole, *I*, to receive the bow

C. The lever B is of the same length as A and has on the upper end at G a coupling made the same as a joint on a buggy top brace. This coupling connects the lever B and bow C. The hole in the coupling and the hole shown at H, Fig. 27, are one inch apart from center to center.

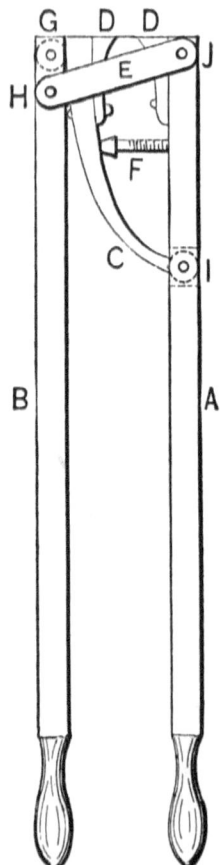

FIG. 26—THE BOLT CUTTER AS COMPLETED.

I next take a piece of steel ¾-inch square to make the bow. I first stave it up on one end to put the ear on it for the coupling G, then I put an eye in it to fit in the long holes shown at I, and bend it so that the knife D will fit closely against it when the two are put together.

To forge the knife I take a piece of ¾-inch good cast-steel. I dress up the knives, harden them and then rivet one on the lever A and the other on the bow C, using two rivets in each one. The plates E, are made of 1 7-8-inch by 3-8-inch iron, the holes in them being four inches apart from center to center.

The holes are 3-8 inch. The plates are placed as indicated by the lines in Fig. 26, and are held in position by steel rivets inserted in the holes H and I. The set screw F is used to prevent the edges of the knives from striking together. The jaws must be open about three-quarters of an inch when the levers are straight. In this tool the cutting is done, not by pressing the levers together but by pulling them apart.

I can cut with it all bolts from ½-inch down. —By L. G.

MAKING A BOLT CLIPPER.

I have made a bolt clipper which, in my opinion, is equal to any of the patent ones in the market. In Fig. 27 of the illustration, A denotes the long handle made of ¾-square iron; B is the other piece, C is the double hinge, D D are the knives, E the purchase lever, F the piece that holds the purchase lever in place.

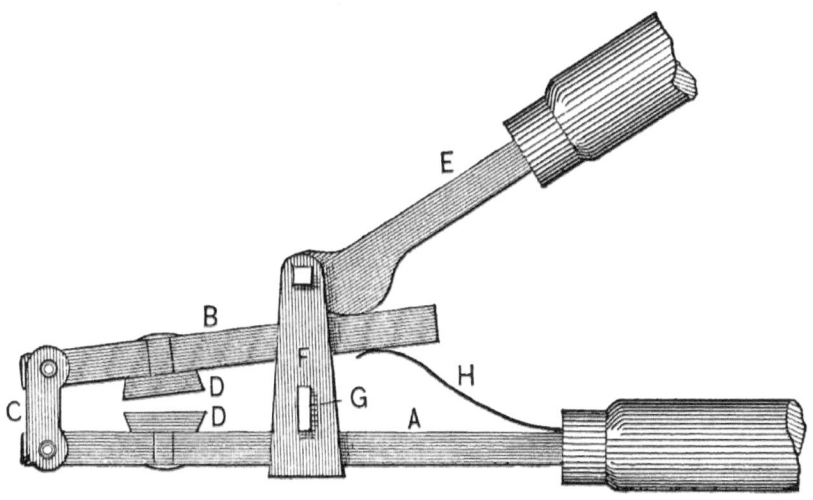

FIG. 27—SIDE VIEW OF BOLT CLIPPER MADE BY "C. N. S."

In Fig. 28 the piece F is shown ready to bend, and in Fig. 29 it is shown bent.

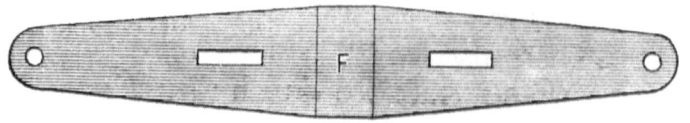

FIG. 28—SHOWING THE PIECE F, USED IN THE BOLT CLIPPER.

G is a key for fastening the piece F on the piece E or the main lever.

FIG. 29—SHOWING THE PIECE F BENT.

It is also used to keep the knives apart. H, in Fig. 27, is the spring used to open the jaws.

FIG. 30—SHOWING ONE OF THE PIECES USED FOR THE HINGE.

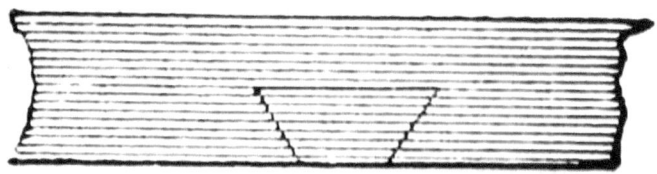

FIG. 31—SHOWING HOW THE KNIVES ARE FASTENED TO THE HANDLE.

The piece shown in Fig. 30 is one of those that form the hinge, one goes on one side and one on the other, being fastened together with two rivets. Fig. 31 shows how the knives are put on the handle. —*By* C. N. S.

TOOL FOR CUTTING RIVETS.

I send you sketches of a pair of cut-nippers, Figs. 32 and 33. They are adapted to cutting bolts and rivets up to ¼-inch in diameter. The jaws do not project, so as to cut long wire, and whatever is cut must be inserted end-ways.

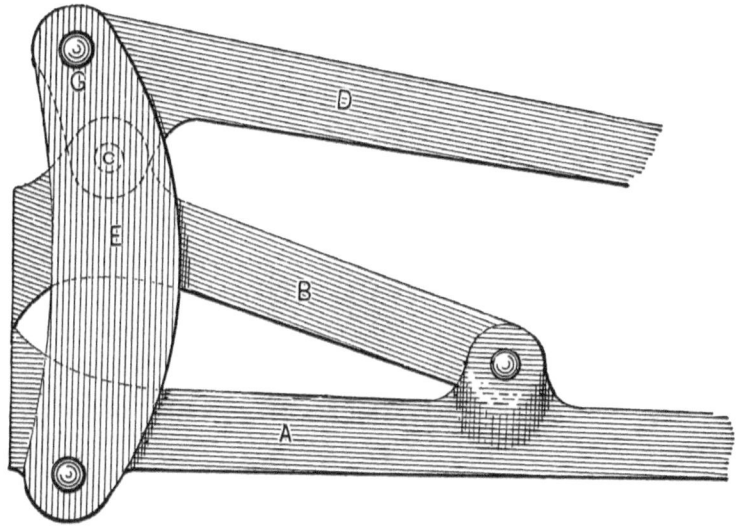

FIG. 32—"O. F. F.'S" RIVET CUTTER.

When finished the tool is 10 inches long, and weighs 14 ounces. *A* is the fixed jaw or leg, having a pivoted jaw, *B*.

A lever, *D*, is pivoted to *B*, at *C*. Two plates, *E*, on each side of the jaws, are pivoted to *A*, at *F*, and to *D*, at *G*; *D* is moved outwards, a rivet, at *R*, put in, and *D* is closed, cutting off the rivet, the operation being obvious. The plate, *E*, must be 1-3 of an inch thick, and riveted to *A* and *D*, on both sides, with at least 3-16 rivets, as the strain is very great. If well made and carefully used, one of these cut-nippers will last a long time. I have used one seven years, and it is in good condition yet, though I have averaged to iron fifty sleighs a year, and they have done all the cutting, besides all my other work, where they could be used. —*By* O. F. F.

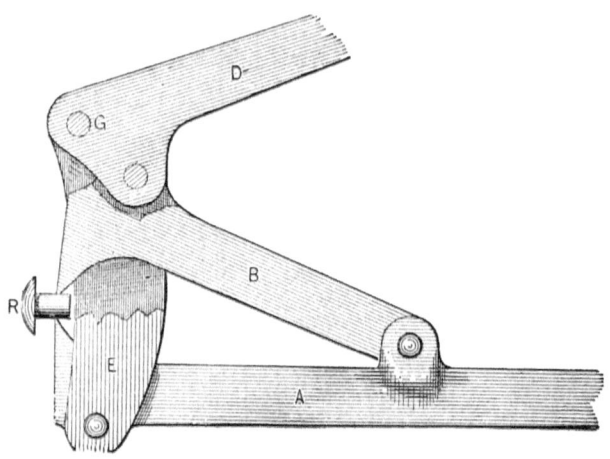

FIG. 33—ANOTHER VIEW OF "O. F. F.'S" RIVET CUTTER.

RIVET CUTTER.

I have a tool which will cut a rivet or bolt one-half inch in diameter very easily. It is very handy and useful in cutting points of bolts, in ironing wagons, buggies, etc.

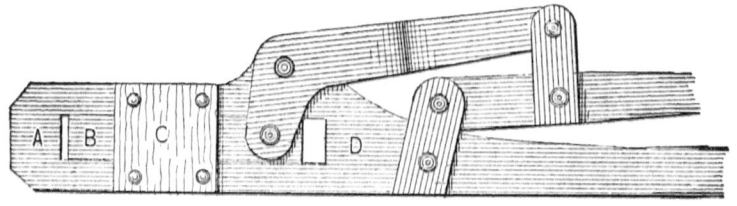

FIG. 34—SHOWING RIVET CUTTER CLOSED

In my sketch, Fig. 34 represents the tool nearly closed. The part marked *A* is one shear or knife, which is a piece of steel (best) welded on the iron frame or body, and beveled from the opposite side so as to make an edge. The part marked *B*, is the main or sliding shear, made of the best steel. It also has a beveled edge the same as *A*.

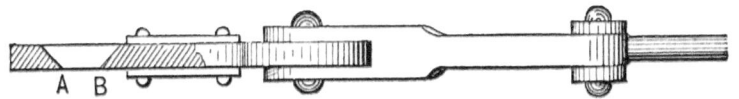

FIG. 35—SECTIONAL VIEW OF RIVET CUTTER SHOWING EDGES OF SHEARS.

C is a plate, and there is one on each side so as to hold the shear or knife, B, to its place. These plates can be fastened on either with rivets or small bolts as desired. D is the main frame or body of the tool, which is iron. Fig. 35 is a top view with a portion of the outer jaw removed, showing the points or edges of shears or knives, A and B, and the method of securing the plates referred to. —*By* Cyrus G. Little.

TOOLS FOR MAKING RIVETS—PIPE TONGS.

I send you a sketch, Fig. 36, of a handy rivet-making tool. The hole at A is just deep enough to make the required length of rivet; the wire is cut off long enough to make the body and head, and is riveted with a button rivet set; the lower part of the tool is bowed as you see and naturally holds the two jaws a little open.

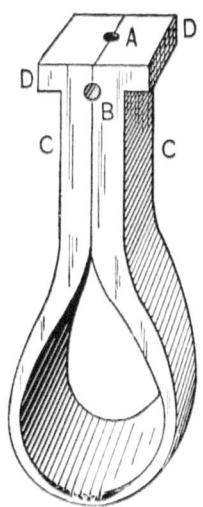

FIG. 36—RIVET-MAKING TOOL.

The vise jaws grip at *C C*, the two flanges, *D*, resting on top of the vise jaws: as the vise is opened or shut the jaws of the tool open and release or close and grip the rivet.

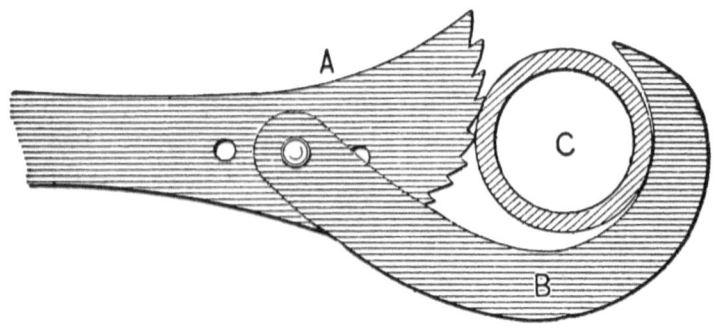

FIG. 37 BEST FORM OF PIPE TONGS.

I also send you a sketch, Fig. 37, of, I think, the best form of pipe tongs. The jaw *B* points to jaw *A*. Moving *A* in the direction of arm causes pipe *C* to be very firmly gripped. —*By* "Southern Blacksmith."

A TOOL FOR MAKING RIVETS.

The accompanying illustration, Fig. 38, represents a tool which is very convenient for making the rivets that are used to fasten the brass on the plow share and the bar, and also the frog when wooden stock plows are made. The tool will make rivets of two lengths, namely, 1 ½-inch and 1-inch. I used 3-8-inch round Norway iron for rivets because it is the only kind fit for that purpose. Rivets made of common iron will always break if they are put in hot.

I make the tool as I would an ordinary heading tool, but am careful to get the ends *A B*, high enough where the holes are. The end *A* is for the 1 ½-inch rivets and the other end is for the 1-inch rivets; *A* is made two inches high, and *B* is an inch and a half high. The ends are laid with steel on the tops, and I then take a 3-8-inch bit the size of the round iron used and bore holes at *C* and *D*, so that they lack but half an inch more to come through the piece, and then bore through the rest of the way with a 3-16-inch bit at *E* and *F*.

FIG. 38—A TOOL DESIGNED BY "L. G." FOR MAKING RIVETS.

This is to facilitate the driving out of the rivets after they are made. The iron should be cut long enough to allow for a head. After making the rivets I drive them out with a small punch. If a little oil is used in the tool they will come out easier. —*By* L. G.

MAKING A BOLT CLIPPER.

I have a bolt clipper that will cut easily bolts of half an inch or smaller ones. It is made as follows: I first make a pattern of tin. For the jaws, which are marked *A A* in the accompanying illustration, Fig. 39, I used a piece of bar iron, 3-8 x 3 inches, cutting off two pieces about 10 inches long, then forming them according to the pattern and welding on a piece of steel for the cutting edge. The hole *B* is made 5-8-inch in diameter. The holes *C C* and *D* are ½-inch in diameter. The distance from the hole *B* to *C C* is 6 inches, from *C C* to *D* it is 1 ¼ inches, and from *C* to *C* is 1 ½ inches. The handles *G G* are made of iron 5-8 x 1 ½ inches and are joined at *D*.

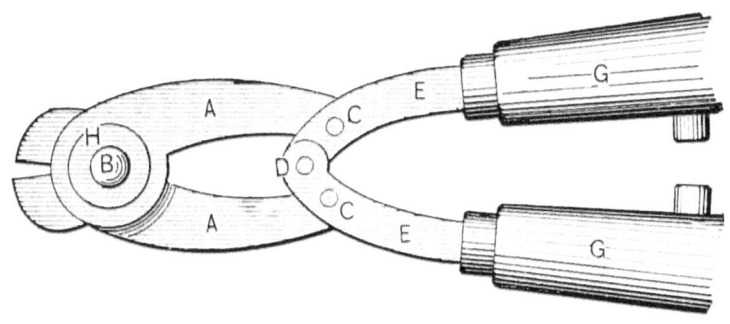

FIG. 39—BOLT CLIPPER MADE BY "W. R."

The jaws, *A A*, are joined to the handles at *C C*. The other parts of the handles, *G G*, are of wood, about three feet long, with ferrules on the ends. The total length of the handles, measuring from *D*, is 3 ½ feet. The washer, *H*, is 2 inches in diameter, and is forged to a thin edge around the outside. I put one on each side. The rivets should be of steel. —*By* W. R.

HOW TO MAKE A BOLT AND RIVET CUTTER.

I have made a bolt and rivet cutter that works splendidly, and will tell how it is made.

Take a piece of square iron, twenty-four inches long, for the bottom piece. In the end weld a piece of tool steel for a cutter. This should be 3-4 inch so as to have solid steel cutters. In the bottom piece, seven inches from the end, punch a hole with a flat punch and round down from above the hole.

The other jaw is made from the same kind of iron with steel welded on the end for cutter.

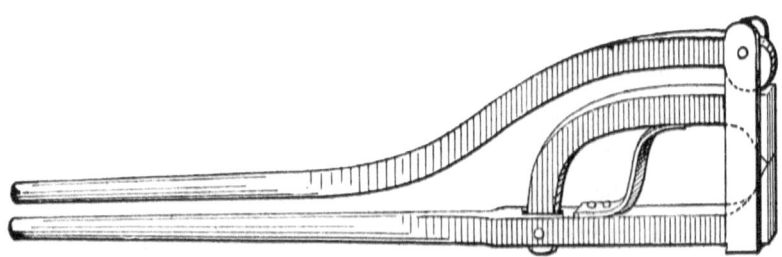

FIG. 41—SHOWING C. V. MARSH'S BOLT AND RIVET CUTTER COMPLETE.

The upper handle can be made from round or square iron whichever is handiest. The spring can be made from any piece of old spring, and is put on with two small rivets. The end piece can be made from 3-4-inch stake iron. It is in one piece and bent so as to fit around the lower jaw. Fig. 41 shows the cutter complete and will give a good idea as to how it is made. I think that some of the boys will find this cutter very useful. It is powerful and will cut easily small bolts and pieces of iron. It is also, as will be seen by the engraving, simple in construction and not difficult for any smith to make. —*By* C. V. Marsh.

CHAPTER III.

CHISELS.

THE CHISEL AND CHISEL-SHAPED TOOLS.

The subject upon which I have been invited by the Franklin Institute to speak this evening is that of the chisel and chisel-shaped tools, and the object of my remarks will be similar to that I had in view in a former lecture, namely, to demonstrate, as far as it is possible in a talk of this kind, that in skillful handicraft the very foundation lies in a knowledge that may be obtained altogether independent of any actual use of the tool.

The first day I entered the machine-shop I was given a hammer and a cold chisel wherewith to chip the ends of some bolts level. I had looked forward to my entry into the shop with a great deal of pleasure, for my heart and mind were set upon becoming a skillful workman. The idea of being able to cut and shape metal to my will, and form it into the machines that were to save mankind the exercise of mere brute force, had such a charm for me that it was the height of my ambition. An apprentice of some two years' standing was to show me how to use the chisel, which he did as follows:

"You hold the chisel so, and the hammer so, and then you chip this way," and he cut off the end of the first bolt very nicely and quickly. I tried to follow him, but after the first blow, which by chance struck the chisel-head sideways, I became aware that my hand was dangerously near to the chisel-head. I realized this more thoroughly at the second blow, for the hammer fell upon my thumb, to the great amusement of my neighbors.

After that I could not be persuaded to hold the chisel near the head unless I held the hammer pretty close to its head, so that I could take better aim. For two days I struggled on, left to myself to find out by bungling along how to grind the chisel, and all the other points that could have been taught me in an hour.

What was worse, I became disheartened, for instead of finding all plain sailing with nothing to do but to master the principles of tool using, feeling every day that I had made some progress, I found myself floundering in the dark, not understanding anything of what I was doing, asking others to grind the chisels because I had no idea how to do it properly myself, and at the end of the first month I should, but for the authority of my parents, have tried some other business. The machinist's trade seemed to be nothing but one-half main strength, one-quarter stupidity, and the other quarter hand skill that every man had to work out for himself, for nobody seemed able to help me.

Many a boy meets just this same experience, and getting discouraged drifts about a month at this trade, two months at that, until he finds himself at last without any trade at all, and very often in his old age without the means of earning an honest livelihood. Examples of this kind are, I believe, within the personal knowledge of most of us, and the fault is often attributed to the absence of an apprenticeship system, but if we go deeper I am persuaded that it will appear that it is more in the want of intelligent preparation for the workshop.

Parental authority, as I have said, saved me from this misfortune, but since then I have, in the course of years, mastered the principle involved in the use of this cold chisel, and I can now draw you two pictures, which I hope will not be uninteresting.

Suppose when I went to the shop doors to ask for employment the superintendent had said to me:

"Want to be a machinist, do you? Well, why do you think you are fitted for it; do you know anything about it, or about tools? On what foundation have you built the opinion that you will ever make a good machinist?"

What could I then have answered except that I thought so, hoped so, and meant to try my best. But suppose I was again a boy, and again found myself at the shop door, having previously taken enough interest in mechanics to have remembered the principles I had already been taught, I could take a pencil and a piece of paper and answer him thus "I can only say, sir, that I have prepared myself somewhat for a trade;" here, for example, in Figs. 42 and 43, are shown the shapes in which flat chisels are made.

The difference between the two is, that the cutting edge should be parallel with the flats on the chisel, and as Fig. 42 has the widest flat, it is easier to tell with it when the cutting edge and the flat are parallel, therefore the broad flat is the best guide in holding the chisel level with the surface to be chipped. Either of these chisels is of a proper width for wrought iron or steel because chisels used on these metals take all the power to drive that can be given with a hammer of the usual proportions for heavy chipping, which is, weight of hammer, 1 3-4 lbs.; length of hammer handle, 13 inches; the handle to be held at its end and swinging back about vertically over the shoulder.

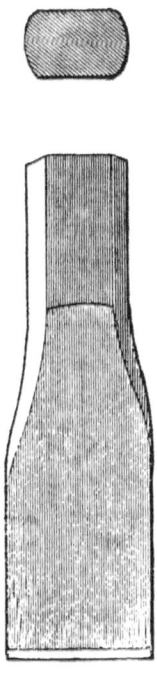

FIG. 42—SHOWING A FLAT CHISEL

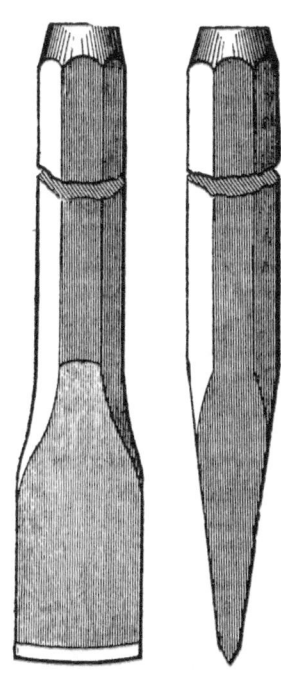

FIG. 43—SHOWING ANOTHER SHAPE OF FLAT CHISEL. SIDE AND END VIEWS.

If I use so narrow a chisel on cast iron or brass, and give full force hammer blows, it will break out the metal instead of cutting it, and the break may come below the depth I want to chip and leave ugly cavities.

So for these metals the chisel must be made broader, as in Fig. 44, so that the force of the blow will be spread over a greater length of chisel edge, and the edge will not move forward so much at each blow, therefore it will not break the metal out.

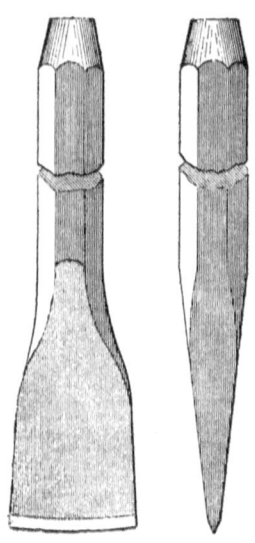

FIG. 44—BROAD CHISELS.

Another advantage is that the broader the chisel the easier it is to hold its edge fair with the work surface and make smooth chipping. The chisel-point I must make as thin as possible, the thickness shown in my sketches being suitable for new chisels.

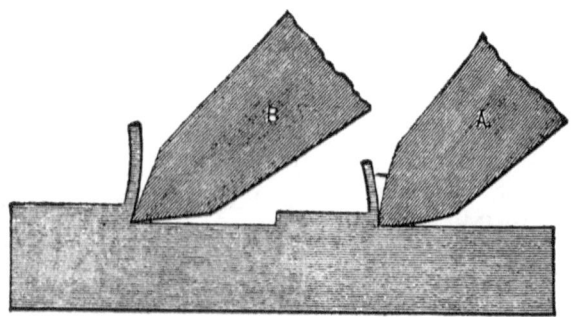

FIG. 45—CORRECTLY AND INCORRECTLY GROUND CHISELS.

In grinding the two facets to form the chisel, I must be careful to avoid grinding them rounded as shown at A in the magnified chisel ends in Fig. 45, the proper way being to grind them flat as at B. I must make the angle of these two facets as acute as I can, because the chisel will then cut easier.

The angle at *C*, in Fig. 46, is about right for brass, and that at *D* is about right for steel. The difference is that with hard metal the more acute angle dulls too quickly.

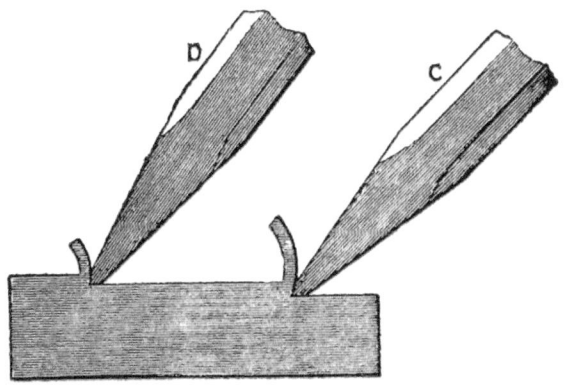

FIG. 46—CHISELS FOR BRASS AND STEEL.

Considering the length of the cutting it may for heavy chipping be made straight as in Fig. 42, or curved as in Fig. 44, which is the best, because the corners are relieved of duty and are therefore less liable to break.

The advantage of the curve is greatest in fine chipping because, as you see in Fig. 47, a thin chip can be taken without cutting with the corners, and these corners are exposed to the eye in keeping the chisel-edge level with the work surface.

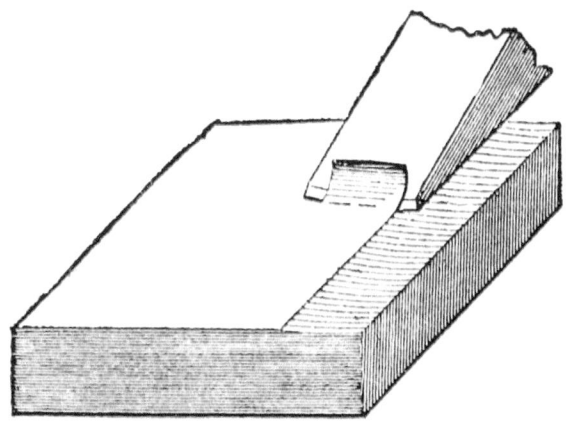

FIG. 47—CHISEL FOR FINE CUTS.

In any case I must not grind it hollow in its length, as in Fig. 48, or as shown exaggerated in Fig. 49, because in that case the corners will dig in and cause the chisel to be beyond my control, and besides that, there will be a force that, acting on the wedge principle and in the direction of the arrows, will operate to spread the corners and break them off.

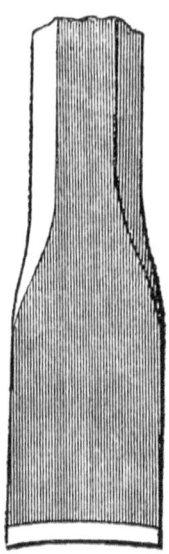

FIG. 48—IMPROPERLY GROUND CHISEL.

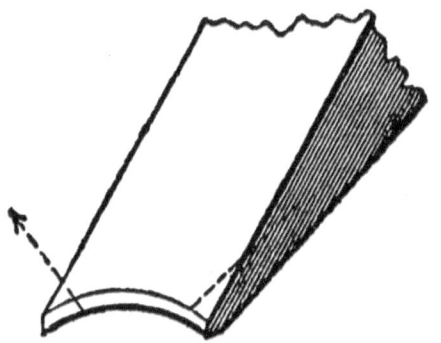

FIG. 49—MAGNIFIED VIEW OF THE CHISEL SHOWN IN FIG. 48.

I must not grind the facets wider on one side than on the other of the chisel, as in Fig. 50, because in that case the flat of the chisel will form no guide to let me know when the cutting edge is level with the work surface.

Nor must I grind it out of square with the chisel body, as in Fig. 51, because in that case the chisel will be apt to jump sideways at each hammer blow.

I can remove a quantity of metal quicker if I use the cape chisel in Fig. 52 to first cut out grooves, as at *A*, *B* and *C*, in Fig. 53, spacing these grooves a little narrower apart than the width of the flat chisel, and thus relieving its corners. I must shape the end of this cape chisel as at *A* and *B*, and not as at *C*, as in Fig. 53, because I want to be able to move it sideways to guide it in a straight line, and the parallel part at *C* will interfere with this, so that if I start the chisel a very little out of line it will go still farther out of line, and I cannot move it sideways to correct this.

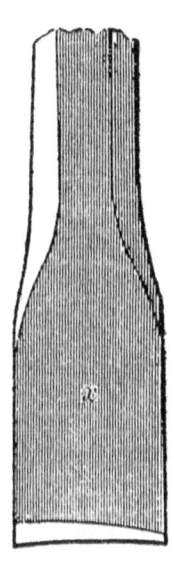 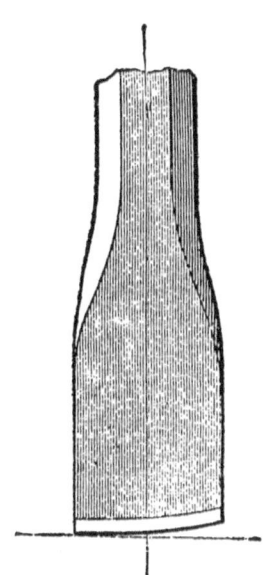

FIG. 50—SHOWING A COMMON ERROR IN GRINDING.

FIG. 51—SHOWING ANOTHER ERROR IN GRINDING.

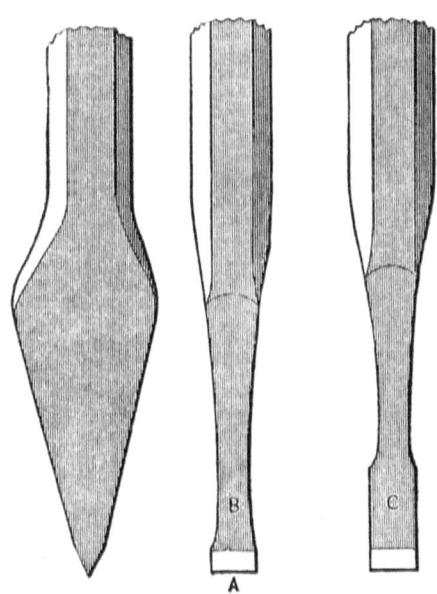

FIG. 52—PROPER AND IMPROPER SHAPES FOR CAPE CHISELS.

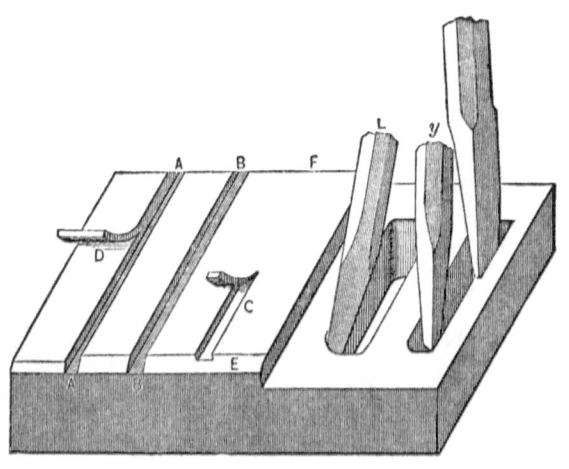

FIG. 53—SHOWING THE APPLICATION OF THE CAPE CHISEL TO FACILITATE THE WORK OF THE FLAT CHISEL.

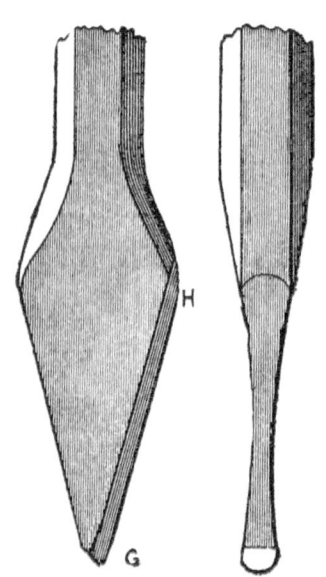 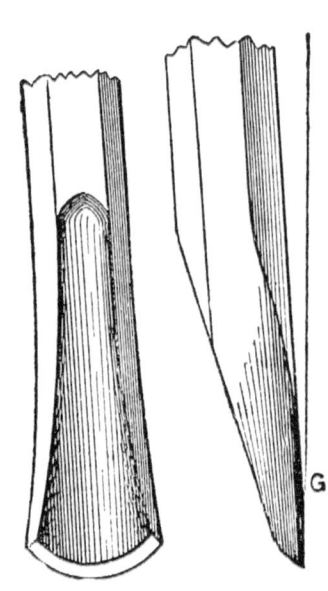

FIG. 54—THE ROUND NOSE CHISEL.

FIG. 55—SHOWING HOW THE COW MOUTH CHISEL IS BEVELED.

The round-nosed chisel, Fig. 53, I must not make straight on its convex edge; it may be straight from *H* to *G*, but from *G* to the point it must be beveled so that by altering the height of the chisel head I can alter the depth of the cut.

The cow-mouthed chisel, Fig. 55, must be beveled in the same way, so that when I use it to cut out a round corner, as at L in Fig. 53, I can move the head to the right or to the left, and thus govern the depth of its cut.

The oil groove chisel in Fig. 56, I must make narrower at A than it is across the curve, as it will wedge in the groove it cuts.

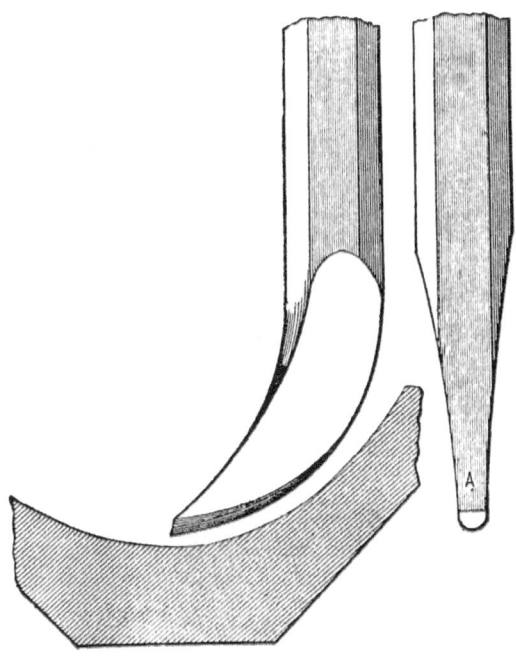

FIG. 56—THE OIL GROOVE CHISEL.

The diamond-point chisel in Figs. 57 and 58 I must shape to suit the work, because if it is not to be used to cut out the corners of very deep holes, I can bevel it at M, and thus bring its point X central to the body of the steel as shown by the dotted line Q, rendering the corner X less liable to break, which is the great trouble with this chisel. But as the bevel at M necessitates the chisel being leaned over as at Y, in Fig. 53, it could, in deep holes, not be kept to its cut; so I must omit the bevel at M, and make that edge straight as at $R\ R$ in Fig. 58.

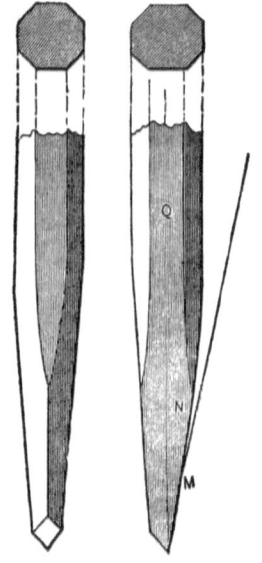 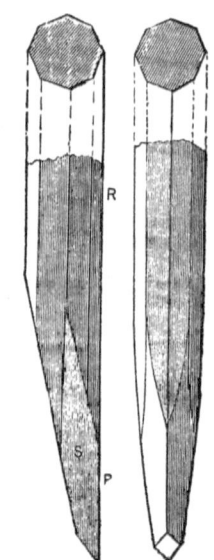

FIG. 57—THE DIAMOND POINT CHISEL FOR SHALLOW WORK.   FIG. 58—THE DIAMOND POINT CHISEL FOR DEEP WORK.

The side chisel obeys just the same rule, so I may give it bevel at *W*, in Fig. 59, for shallow holes and lean it over as at *Z* in Fig. 53 or make the side *V W* straight along its whole length, for deep ones; but in all chisels for slots or mortises it is desirable to have, if the circumstances will permit, some bevel on the side that meets the work, so that the depth of the cut can be regulated by moving the chisel head.

In all these chisels, the chip on the work steadies the cutting end, and it is clear that the nearer I hold the chisel at its head the steadier I can hold it and the less the liability to hit my fingers, while the chipped surface will be smoother.

Now, what I have said here is what I might have learned before I applied at the shop, and is it not almost a certainty that if there was a vacancy I should have obtained the position? Nay, more, I venture to say that I should have received the appointment before I had made half my explanation, unless, indeed, the superintendent heard me through out of mere curiosity, for it certainly would, as things now are, be a curiosity for a boy to have any idea of the principle involved in using tools before he had them actually placed in his hands—unless, indeed, it be the surgeon's tools.

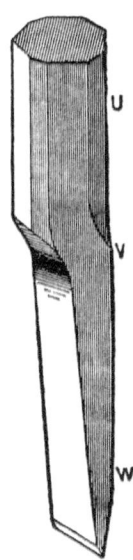

FIG. 59—THE SIDE CHISEL.

There is an old saying that an ounce of practice is worth a pound of theory, but this sounds to me very much like saying that we should do a thing first and find out how it ought to be done afterwards. Yet I should not care to patronize a young dentist or a young surgeon who was pursuing his profession in this way.

I may, however, illustrate to you some of the points I have explained by adding to the pound of theory I have advanced an ounce of practice. Here, for example, I have to take a chip off a piece of wrought-iron, and, as it is a heavy chip, I stand well away from the vise, as an old hand would do, instead of close to it, as would be natural in an uninstructed beginner. In the one case you will observe that the body is lithe and supple, having a slight motion in unison with the hammer, while in the other it is constrained, and not only feels but looks awkward. If, now, I wish to take a light chip, I must stand nearer to the work, so that I can watch the chisel's action and keep its depth of cut level. In both cases I push the chisel forward to its cut and hold it as steadily as I can.

It is a mistake to move it at each blow in this way, as many do, because it cannot be so accurately maintained at the proper weight.

Here I take a deep cut on a piece of brass, and the full force blows have broken it out, for the reasons I explained just now. Next we will take a finishing cut across, leaving the surface smooth and more level for the filing that is to follow.

Light and quick blows are always necessary for the finishing cuts, whatever the kind of metal may be.

Here are two cape chisels, one formed as at B and the other as at C, in Fig. 52, and a cut being taken with each, you will see that I have been able to direct the path sideways of B, but that I could not do so with C.

With the side chisel alone I can illustrate the points made with reference to the chisel shown in

Figs. 54, 55, 57, 58 and 59, namely, that there must be a bevel made at the end in order to enable the depth of cut to be adjusted and governed, for if I happened to get the straight chisel too deeply into its cut I cannot alter it, and unless I begin a new cut it will get imbedded deeper and will finally break. But with this side chisel, Fig. 59, that is slightly beveled, I can regulate the depth of cut, making it less if it gets too deep, or deeper if it gets too shallow.

The chisel that is driven by hammer blows may be said to be to some extent a connecting link between the hammer and the cutting tool, the main difference being that the chisel moves to the work while the work generally moves to the cutting tool. In many stone-dressing tools the chisel and hammer are combined, inasmuch as that the end of the hammer is chisel-shaped, an example of this kind of tool being given in the pick that flour millers use to dress their grinding stones. On the other hand we may show the connection between the chisel and the cutting tool by the fact that the wood-worker uses the chisel by driving it with a mallet, and also by using it for a cutting tool for work driven in the lathe. Indeed, we may take one of these carpenter's chisels and fasten it to the revolving shaft of a wood-planing machine, and it becomes a planing knife; or we may put it into a carpenter's hand plane, and by pushing it to the work it becomes a plane blade. In each case it is simply a wedge whose end is made more or less acute so as to make it as sharp as possible, while still retaining strength enough to sever the material it is to operate upon.

In whatever form we may apply this wedge, there are certain well-defined mechanical principles that govern its use.

Thus when we employ it as a hand tool its direction of motion under hammer blows is governed by the inclination of that of its faces which meets the strongest side of the work, while it is the weakest side of the material that moves the most to admit the wedge and therefore becomes the chip, cutting, or shaving. In Fig. 60, for example, we have the carpenter's chisel operating at *A* and *B* to cut out a recess or mortise, and it is seen that so long as the face of the chisel that is next to the work is placed level with the straight surface of the work the depth of cut will be equal, or, in other words, the line of motion of the chisel is that of the chisel face that lies against the work.

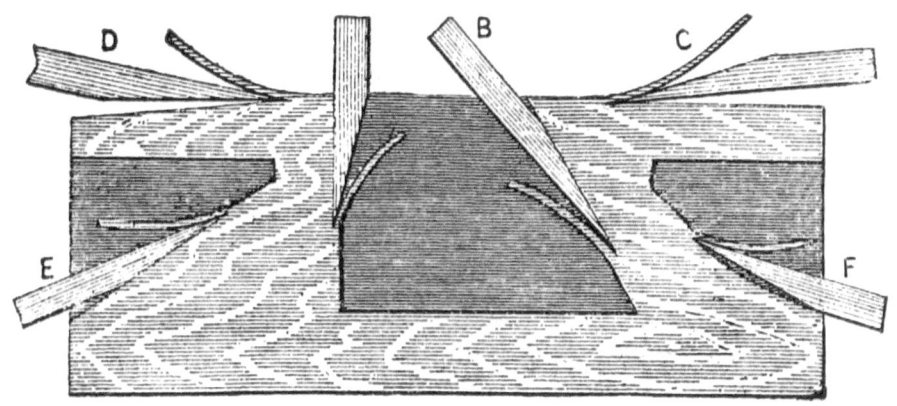

FIG. 60—SHOWING THAT THE DEPTH OF THE CUT DEPENDS UPON THE POSITION AND DIRECTION OF THE LOWER SURFACE OF THE CHISEL.

At *C* and *D* is a chisel with, in the one instance, the straight, and in the other, the beveled face toward the work surface. In both cases the cut would gradually deepen because the lower surface of the chisel is not parallel to the face of the work.

If now we consider the extreme cutting edge of chisel or wedge-shaped tools it will readily occur that but for the metal behind this fine edge the shaving or cutting would come off in a straight ribbon and that the bend or curl that the cutting assumes increases with the angle of the face of the wedge that meets the cutting, shaving or chip.

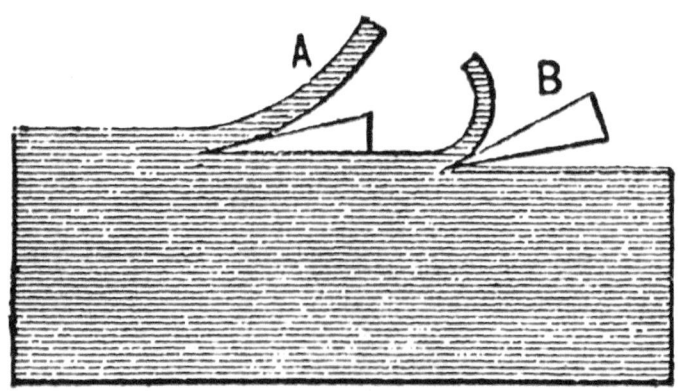

FIG. 61—SHOWING THAT THE EFFECT OF THE CUTTING EDGE DEPENDS UPON THE ANGLE OF THE BOTTOM FACE TO THE CHISEL'S LINE OF MOTION.

I may, for example, take a piece of lead and with a pen-knife held as at *A*, Fig. 61, cut off a curl bent to a large curve, but if I hold the same knife as at *B* it will cause the shaving to curl up more. Now it has taken some power to effect this extra bending or curling, and it is therefore desirable to avoid it as far as possible. For the purpose of distinction we may call that face of the chisel which meets the shaving the top face, and that which lies next to the main body of the work the bottom face.

Now at whatever angle these two faces of the chisel may be to the other and in whatever way we present the chisel to the work, the strength of the cutting edge depends upon the angle of the bottom face to the line of motion of the chisel, and this is a principle that applies to all tools embodying the wedge principle, whether they are moved by a machine or by hand.

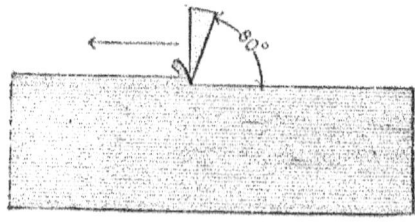

FIG. 62—SHOWING THE BOTTOM FACE AT AN ANGLE OF 80 DEGREES TO THE LINE OF MOTION

Thus, in Fig. 62, we have placed the bottom face at an angle of 80 degrees to the line of tool motion, which is denoted by the arrow, and we at once perceive its weakness. If the angle of the top face to the line of tool motion is determined upon, we may therefore obtain the strongest cutting edge in a hand-moved tool by causing the bottom angle to lie flat upon the work surface.

But in tools driven by power, and therefore accurately guided in their line of motion, it is preferable to let the bottom face clear the work surface, save at the extreme cutting edge.

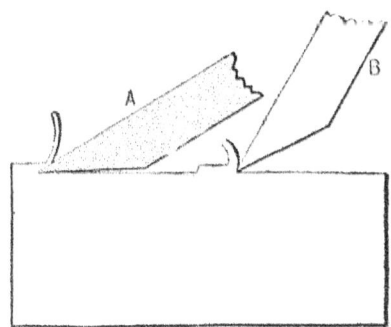

FIG. 63—SHOWING TWO POSITIONS OF THE WEDGE.

The front face of the wedge or tool is that which mainly determines its keenness, as may be seen from Fig. 63, in which we have the wedge or tool differently placed with relation to the work, that in position A obviously being the keenest and least liable to break from the strain of the cutting process. — *From a lecture delivered by* Joshua Rose *before the Franklin Institute, Philadelphia.*

CHIPPING AND COLD CHISELS.

Permit me to make some remarks on my experience with chipping chisels.

"There's not much of interest in the subject," you may say, "for everybody knows all about cold chisels."

Not exactly, for there are a good many chisels that are not properly shaped. Figs. 64 and 65 represent common shapes of cape chisels. That in Fig. 64 is faulty because it is a parallel or nearly so from *A* to *B* and a straight taper from *B* to *C*; its being parallel from *A* to *B* causes it to stick and jam in the groove it cuts, or even to wedge when the corners of the cutting edge get a little dulled; while if they should break (and these corners sometimes do break) there is the whole of the flat place to grind, if the side is ground at all, as it is desirable when the break extends up the chisel and not across its cutting edge.

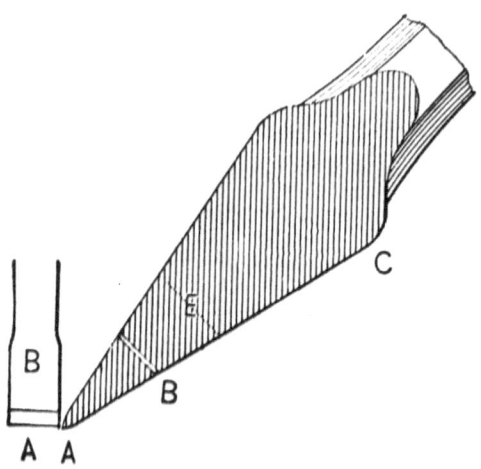

FIG. 64—CHIPPING AND COLD CHISELS. A CHISEL FAULTY AT THE POINT.

"The sticking don't amount to much nor does the grinding," is the answer.

It amounts to some unnecessary sticking that makes it very difficult to alter the angle of the chisel if it is going too deep or not deep enough, and so it is an impediment to smooth, even chipping. The grinding amounts to some unnecessary grinding, and furthermore, the chisel thus shaped is more difficult to forge, very little more difficult I grant, but more difficult all the same.

Haven't you seen men tug at a chisel to get it out of a keyway? Haven't you seen them hit it sideways with a hammer to loosen it in the sides of the cut? I have.

Fig. 64 would do very well for a keyway in a bore, but for outside work it is also faulty because it is too weak across E; hence Fig. 65 is, for outside work, the best shape, being stiffer and therefore less springy.

All these I think are plain and well grounded points, and so to settle a discussion on them I was blindfolded and given three cape chisels, two like Fig. 65, and one like Fig. 64, and in a dozen trials at chipping told each time I was given the one like Fig. 64. I claim that the shape makes a tangible difference. I could tell by the chipping, for it was a piece of machine steel I was chipping, and the corners of Fig. 64 soon began to round and the chisel to wedge.

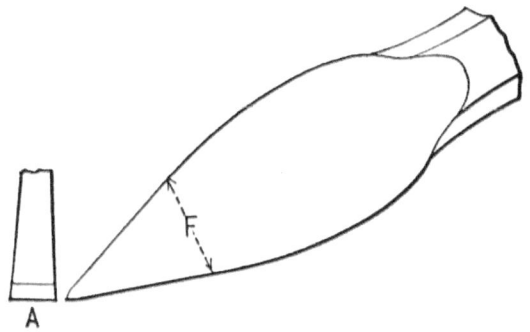

FIG. 65—A BETTER SHAPE FOR COLD CHISELS.

As to the flat chisel, haven't you often seen it hollow along the cutting edge, and isn't that more likely to break and more liable to stick than one a little rounding?

There is one more point that I will mention, and that is a habit many have of pulling the flat chisel back from the cut after every blow. I have seen some good workmen do it, and I am not disposed to find particular fault with it, but I think it is unnecessary, at least I see no end that it accomplishes. I like the chisel to lie steadily under a little hand pressure against the cut so that I can feel that the lower face of the chisel rests fairly and evenly upon the bottom face (as it must do to chip straight), and having got it at the proper angle to the work, I like to carry the cut clear across without moving it once. It is a kind of machine chipping that reminds one of Rowell's running, on, on, on; it goes without a falter. Now a word about using the hammer, not that there is much to discuss about it, but simply to round off the subject. The old style was a 1 3-4 lb. hammer with handle 15 inches long, and this is all right for the man who does chipping enough to keep his muscles well hardened and can swing his hammer ten hours a day without feeling it next day, but it is better to get broke in with a 1 1-4 lb. hammer.

I had at one time a 1 3-4 lb. hammer and a 1 1-4 (or a little heavier than that) hammer, and was well broke in at chipping, having had about a year at getting out work with hammer, chisel and file; the 1 3-4 hammer broke and I took to the smaller one; I found that I could not do as much work with it and it began to tell on my hands, because I could use the lighter hammer quicker, and in doing this I naturally gripped it tighter and it told on me, indeed it would sometimes be a minute before I could straighten my fingers out after releasing the hammer; of course the handle was a little less in diameter, and that had something to do with it, but not all, because my left hand, gripping the chisel, which I always did firmly, never tired, nor did the fingers stiffen, though the diameter of the chisel was smaller than the small hammer handle.

The palm of my hand would, with the small hammer, get red and feel nervous and twitchy, while it would not do so with the heavy one.

Did you ever notice what different styles there are in using a chipping hammer, how much more the wrist and elbow are used by some than others? Yes, and there are graceful and ungraceful chippers. I like to see (I am talking of heavy hand chipping of course) the chipper stand, for the heavy cuts, not too close to the vise, use the wrist very little, the elbow not much, the shoulder a good deal, and to let the body swing a little with the hammer, the ham-mer head going as far back as about vertically over the shoulder, and that is the time when every blow tells.

For the last or smoothing cut I like to see the hammer handle held a quarter way up from the end, the chipper to stand pretty close to his work, using the wrist a little and the elbow, but not the shoulder joint; and just in proportion as the chisels are smaller, the wrist used more and the elbow less.

"What is a good day's chipping for a man thoroughly broken in?"

Well, I should say on a chipping strip of cast iron 3-4 inch wide, taking a cut, say, a full 1-16 inch deep, 600 running inches is a good day's work, to keep it up day after day. —*By* Hammer and Tongs.

HOW TO MAKE COLD CHISELS.

What I have to say about cold chisels is from purely practical experience. In the first place we do not get the best quality of cast steel, and the kind we do get is very inferior to what we used to have in years gone by (say "befo' de wa'"). We now use what is called the Black Diamond, and this is not often very suitable for heavy cutting such as steel rails. When a chisel comes from the hammer of the smith, as a general rule, it is taken to the grindstone and given a bevel and then it is called ready for use.

But it is not. If a chisel is made, tempered and ground properly it will stand until the head wears down to the eyes. In Fig. 66 of the accompanying illustrations, the reader can readily see that the chisel is not true with the hammer marks on each side, and that it also has hammer marks on the edges when it is made and tempered. It may seem as if this would not make a great difference, but nevertheless it does.

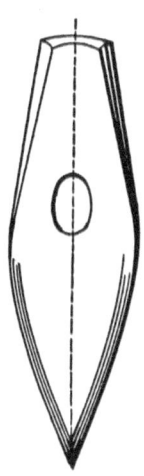
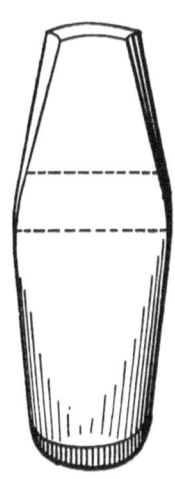

FIG. 66—SHOWING A FAULTY METHOD OF MAKING A CHISEL.

FIG. 67—SHOWING HOW THE HAMMER MARKS ARE GROUND OUT.

FIG. 68—SHOWING HOW THE CHISEL IS BEVELED.

When I make a chisel and temper it (I have to find the proper temper to put in the steel I am working, as steel differs in grade), I take it to the grinding stone and grind out the hammer marks on each side half way up to the eye, and on edges as seen in Fig. 67, so that it will be in the center of the chisel represented by dotted line in Fig. 67. By grinding all the hammer marks out on each side the tool becomes less liable to jar or chatter, and it is jarring or chattering which generally causes the chisel to break. I grind it rounding from the eye or half way from the eye to the point on both sides, after which I give it the bevel as in Fig. 68. I round it on the sides as an axe is rounded and also on the edge.

Cast steel should be worked with charcoal, which adds to instead of diminishing the most important element in steel, which is carbon, while stone coal, through its sulphur, takes away the carbon. The continual use of stone for smith coal reduces the steel and makes it almost worthless for tools. —*By* W.

FORGING COLD CHISELS.

Many blacksmiths find a difficulty in drawing out a cold chisel so that it will cut steel, but if they forge their chisels as I do, the trouble will disappear. Thus, let Fig. 69 represent the chisel to be drawn out to the dotted lines.

FIG. 69—CHISEL TO BE DRAWN OUT.   FIG. 70—SHOWS SHAPE AFTER FORGING

Heat the chisel about as far up as shown in the cut to a blood red, and first forge it down to the dotted lines in Fig. 70; then flatten out the sides, but do not hammer the steel after it is cooled below a red heat. Strike quick and not too heavy blows, especially on the edge. —*By* O. P.

CHAPTER IV.

DRILLS AND DRILLING.

MAKING A DRILL PRESS.

The drill press shown in the accompanying engraving, if properly made, will drill a perfectly straight hole. It will drill from the smallest to the largest hole without any danger of breaking the bit.

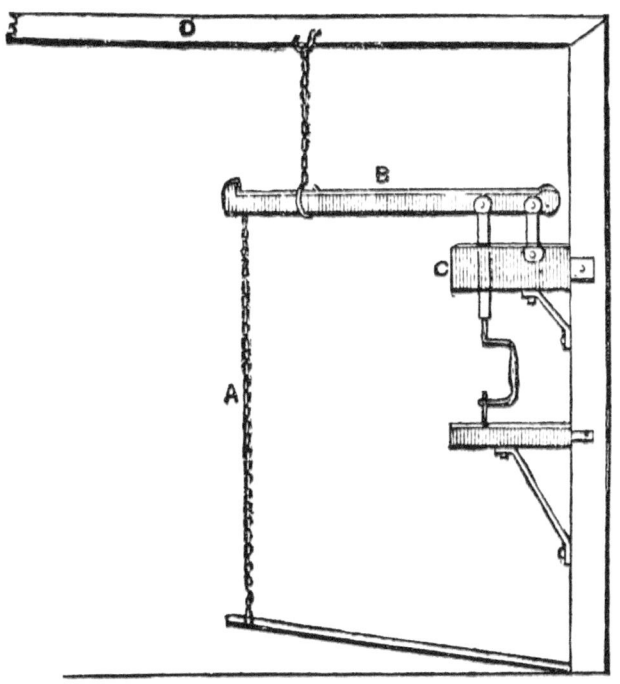

FIG. 71—A DRILL PRESS AS DESIGNED BY "R. E."

The lever B, in Fig. 71, is 1 ¼ x ½ inch and five feet long. The part C is 3 x 6 inches. The chain A can be of the length most convenient for the operator. —*By* R. E.

DRILLING IN THE BLACKSMITH SHOP.

In my shop there is not much drilling done, and what there is done is composed principally of small holes and countersinking for holes ¼ inch and less in diameter. For this purpose I use an ordinary hand drill, but I have found, as I expect other people have found also, that for larger holes than those above named the breast drill becomes quite a nuisance. It is almost impossible to hold it steady, and it drives entirely too hard. It takes a great deal of pushing to get the drill to feed, no matter how thin the drill point is made. So, as I said above, I discarded the breast drill for all holes over ¼ inch in diameter. For holes from ¼ inch up to about ½ inch I used the clamp-shaped rest shown in Fig. 72. I made the back of *A* very much broader than is generally done in such cases, so as to prevent it from bending, a great fault in articles of this kind, as commonly made. I made the work table, *T*, about 6 inches square and gave the feed screw, *S*, a much finer pitch than is usual. I employ 12 threads per inch, by means of which I can feed as lightly as I like, I use all square shank drills, and let the end of the shank pass through the socket so that I can knock the drill out easily.

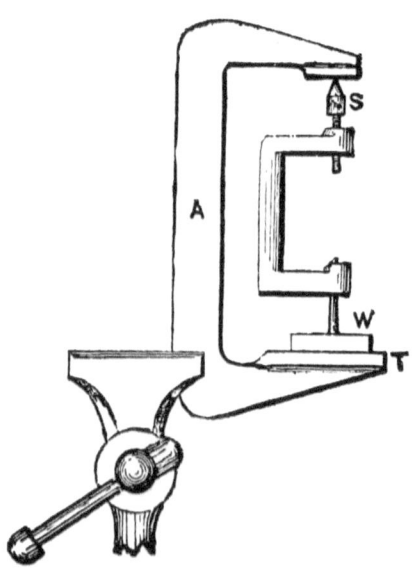

FIG. 72—AN ORDINARY HAND DRILL IMPROVED.

This answers very well for holes that are not too deep. A man can stand the work in such cases, For deep holes and those from ½ inch in diameter upwards I have a small machine, shown in Fig. 73, which I had made to order. The reader will see that it has no self-feed on it.

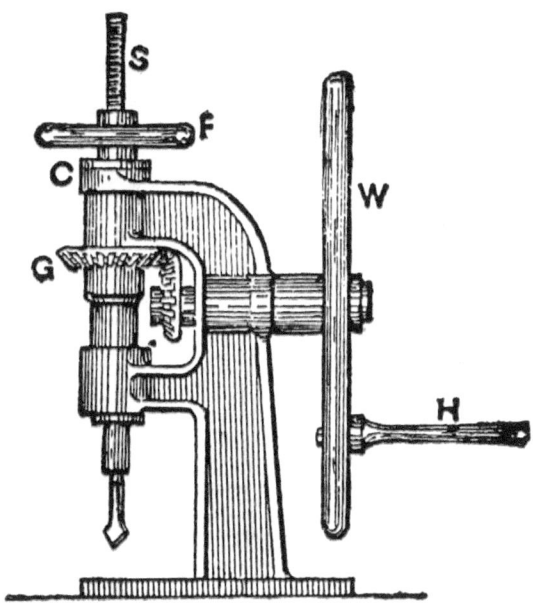

FIG. 73—A BENCH DRILL.

Inasmuch as the operator has got to stand by the machine and mind the wheel it is just as easy, it seems to me, to put on the feed, and in doing so he can increase or diminish it according to the feel of the drill. In this case, as in the other already mentioned, I made the feed screw, S, with a pitch of 12 threads to the inch, and bushed the feed wheel, F, so that I could put in a new nut whenever the threads wore. I made the cap, C, screw on so that I could easily take it off and put in a washer for taking up any lost motion in the feed screw collar. I bought cut gear, G, for I think every drill should have such gear, because they run so much easier. The drilling takes enough hard work to drive without losing any power through cast gear wheels. I made the driving wheel, W, larger than usual, and fastened the handle, H, in the slot so that I could move it further away or closer to the hub, according to circumstances. Every one of the alterations from the ordinary form here described, I am convinced, are substantial improvements.

In the course of time work came along that I could not get under the machine last mentioned, and other work came in that required holes too big to be bored in a hand machine at all. For them I had to resort to the ratchet brace. The rig for the ratchet brace I found, as I presume every other one has found who does an odd job, to be a complete nuisance; still, I had no choice. One day, I put a new hand, a repairer I had hired, on a ratchet-brace job, and left him at it, going away from my business for two days' time to see about some other work. When I got back I found he had rigged up what he called the blacksmith's drill-frame, and which he said was common enough in Scotland, but of which I only knew of one other in this country. I did not like the look of the thing. It appeared like a cross between a gallows frame and some sort of a weighing machine. However, I did not say a word to my man, because I felt a degree of uncertainty about the matter. It might be all right, and so I waited developments. I asked him:

"What, will it drill any better than can be drilled by the method we have formerly used?" He made reply:

"It will drill anything you can get between the posts, and from ¼-inch hole up to a 2-inch hole or more."

He proved this assertion by drilling first a ¼-inch hole, and then a 1 ½-inch hole, that being the largest drill at hand. For the large hole he used the ratchet brace, using the frame as a feeding fulcrum. Fig. 74, of my sketches, represents the device in question. It has two posts, A, and two posts, B, fast to the floor and ceiling. The fulcrum lever is pivoted at C, and has a feeding weight at the opposite end. The lifting lever is pivoted at D, and has, at F, a link connected to the end of the fulcrum lever. At E is an iron plate for the drill brace to rest against, and G is a handle to operate the lifting lever. The work is rested in a movable, or it may be a fixed bench.

The one in question, however, is made movable. By pulling the handle, G, the fulcrum lever is raised, and the drill brace or ratchet brace may be put in position on the work. When G is released the weight pulls down the fulcrum lever to feed the drill brace to its cut. If the weight is too heavy the pressure may be relieved by pulling upon G, or by moving the work further from the posts *A*, the pressure becomes less, because the leverage of the weight is less. This device has one fault, which is that as the fulcrum lever descends in the arc of a circle, as indicated by *H*, it may, in deep holes, become necessary to, move the upper end of the brace to drill the hole straight in the work. The fulcrum lever may be raised or lowered for different heights of work by shifting the pin *C* higher, there being holes at *A*, at intervals, for that purpose.

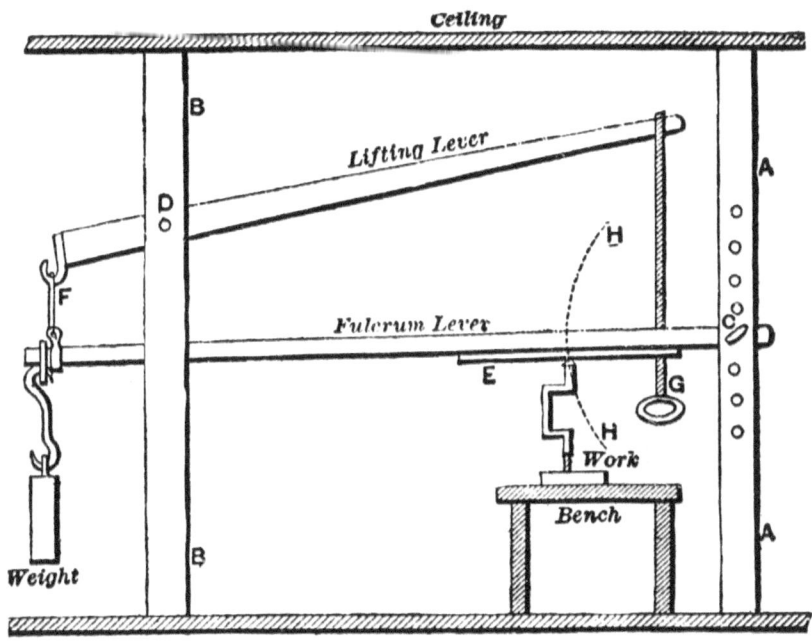

FIG. 74—THE DEVICE INVENTED BY "BLACKSMITH'S" MAN.

This device may be a very old one. It is certainly good for the purpose, however, and very desirable for use where there is no power drill-machine. I would not be without it for many times its cost. —*By* Blacksmith.

A SIMPLE DRILL PRESS.

I send, as shown in the accompanying engraving, Fig. 75, a rig which is simple and also avoids the "arc" direction which has been complained of. A common iron bench screw is inserted in a 4x4 scantling, mortised, over the bench, into upright posts. The cut explains itself. The screw (two feet long) may be had for $1.00, or the smith may make one himself, having a hand wheel. This arrangement takes no room. —*By* Will Tod.

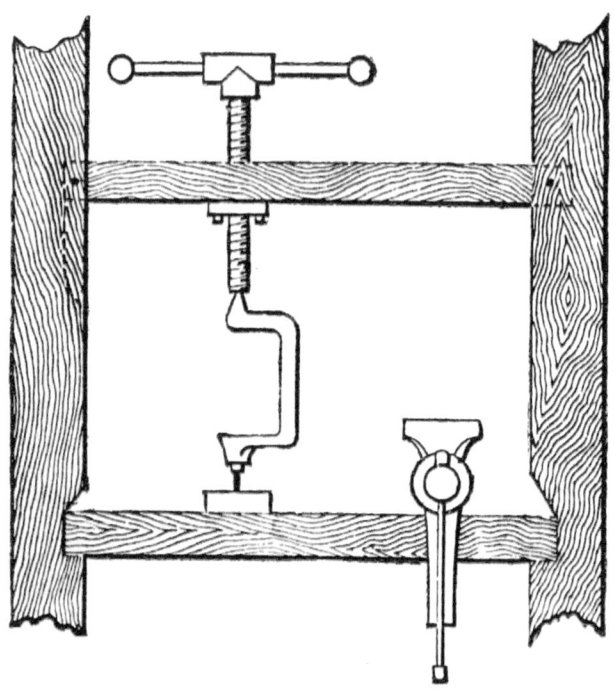

FIG. 75—SIMPLE DRILL PRESS, AS MADE BY WILL TOD.

MAKING A SMALL DRILL.

A very serviceable drill may be made by welding the socket of a shoemaker's awl into a 3-8-inch rod, 5 inches long, with a countersink at the upper end forming a cup to hold lubricating oil and in which the conical center of the feed screw can work. In the engraving, Fig. 76, A represents the screw and B the spindle. Bore a hole through a block of wood to receive the center of the spindle and put the spindle in a two-centered lathe. Move it with a "dog" and a turn pulley like a common thread spool. The drill is run with a bow C, holding all in a vise. Tempered awl blades make good drill bits. —By C. W. D.

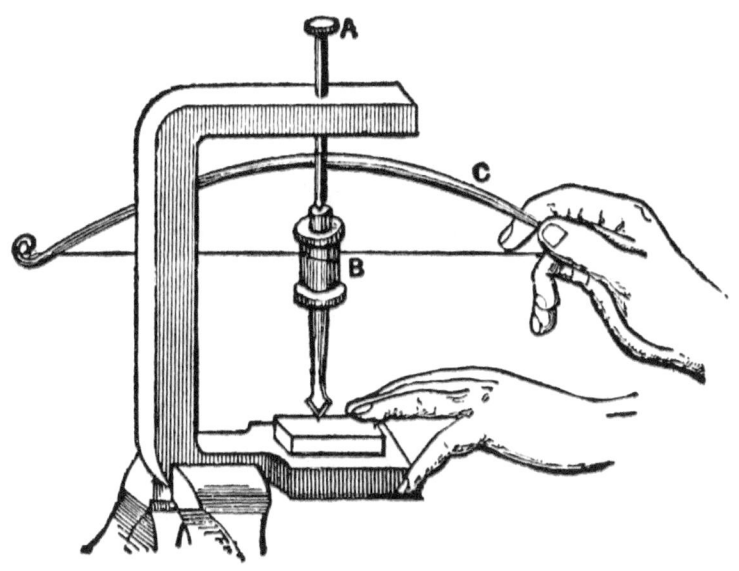

FIG. 76—A SMALL DRILL, AS MADE BY "C. W. D."

TO DRILL A CHILLED MOLD-BOARD.

If you want to drill a hole or file a notch in a stove plate or ploughshare, or other piece of cast-iron, lay it on the fire level until it is cherry red, and then with tongs lay a bit of brimstone on the spot you wish to soften, the piece of brimstone being a trifle less in diameter than the hole you need. Leave the iron on the fire until cold enough to handle and it will yield to your tools. —*By* D. T.

HOLDING LONG BARS IN DRILLING.

A good method of holding long bars of iron, such as sled shoes, so that the holes can be drilled in them easily by one person, is as follows: Take a strong ¼-inch cord or rope and fasten it to the ceiling about five, six or seven feet from the drilling machine, then fasten a pound nut on the end of the rope and let it reach nearly to the floor. When you wish to drill iron, wrap the rope around the iron once at the height you want and you will find that you will have a very handy tool. You can drive a nail so as to hang it up out of the way when not in use. —By A. W. B.

DRILLING GLASS.

Stick a piece of stiff clay or putty on the part where you wish to make the hole. Make a hole in the putty the size you want the hole, reaching to the glass, of course. Into this hole pour a little molten lead, when, unless it is very thick glass, the piece will immediately drop out.

STRAIGHTENING SHAFTS OR SCREWS—A REMEDY FOR DULL AND SQUEAKING DRILLS.

Every machinist who has ever attempted to straighten a polished shaft or screw knows the difficulty of marking the point of untruth when the work is revolved on the centers of a lathe. By procuring a piece of copper pointed on one end and of a shape suitable to fill the tool post, and allowing it to touch the work as it turns, a red mark will be left, even on a brightly polished surface, and this will furnish the desired guide for correction, and at the same time if a short piece of octagon steel about 1 ¾ inches diameter is allowed to partly lie beside the tool post in its T slot, the straightening bar used may be fulcrumed on this with the copper tool still remaining in the tool post, thus expediting the work. If a piece of sheet copper is screwed or riveted to the end of the bar used for straightening, no injury will result to the work from its contact therewith.

Blacksmiths and machinists who use twist drills, have probably been bothered when drilling moderately hard steel by the squeaking and slow cutting of the drill caused by the rapid dulling of its edges. To remedy this, first sharpen the drill, then procure a small piece of tool steel, say 4 inches long, half an inch wide, and 3-16 to ¼ of an inch thick, and after rounding and tempering one end, place the offending drill in a vise, between a pair of copper clamps, gripping it so that the cutting points will be well supported. Then by holding the tempered point of the tool I have described, against the lips of the drill and striking lightly with a hammer on the opposite end, the lips will be upset so that a good clearance will be secured, and the results will be satisfactory if the operation has been carefully done.

This upsetting will, of course, slightly enlarge the diameter of the drill, but in most cases this will do no harm. —*By* J. F. Ross.

A CHINESE DRILL.

Some time ago I read an account of the high quality of Chinese steel. I think there must be some mistake about it. During five years' residence in China, I often examined and remarked the inferior quality of their drills, gravers, etc., and I think their best steel is all imported from England, as I know their finer iron is.

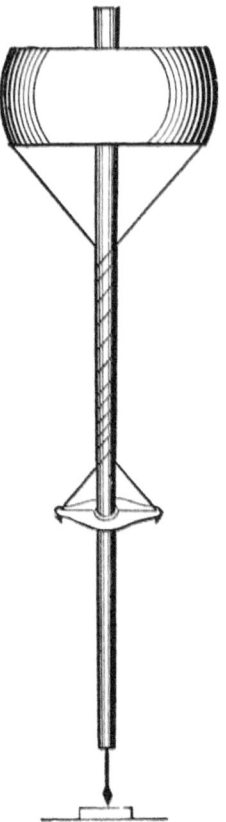

FIG. 77—A CHINESE DRILL AS DESCRIBED BY WILL TOD.

They have an ingenious arrangement for drilling, which is remarkably rapid. As shown in the accompanying engraving, Fig. 77, the drill is fixed in an upright bamboo, which is weighted by a stone (not unlike an old grindstone), at top. It is attached by strings from the top to a handpiece which slides up and down the lower end of the rod.

When the rod is revolved and the hand piece held still, the strings wind on the rod and raise the hand piece, and the machine is "wound up." To start drilling, press the hand piece down the rod till the strings become unwound by the rod revolving; lighten the pressure and the momentum will wind the machine up the reverse way, when pressure is again resumed. —*By* Will Tod.

A DRILL AND COUNTERSINK COMBINED.

I enclose sketch, Fig. 78, of a tool made by myself last Summer, and which may be of some interest to carriage-smiths and blacksmiths who do tiring.

It is a drill and countersink combined, for use on buggy tires. It can also be applied to drilling and countersinking sleigh and sled shoes. It makes quite a saving in time by removing the necessity for using a drill and a countersink separately.

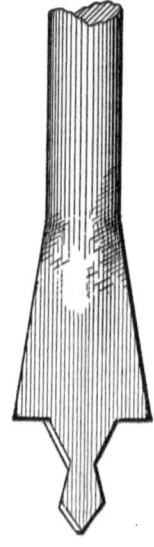

FIG. 78—A DRILL AND COUNTERSINK, AS MADE BY C. H. PREBLE.

It is made so that when the drill part begins to go through the tire, the countersink begins to cut, and when the work is countersunk to the proper depth it is stopped by a shoulder on the tool. Any blacksmith can make this tool in a very short time, and after using one he will never go back to the old method. —*By* C. H. Preble.

A HANDY DRILL.

I have a drill made by myself that is simple, strong, and very effective. Any blacksmith can make it in the following manner:

Take a two-inch rod of iron long enough to reach from the bench to the shop loft, then take two bars of iron, 1 ½ by 5-8 inches, and turn good solid eyes on them, as shown in Fig. 79 of the illustration.

FIG. 79—SHOWING THE BARS WITH EYES TURNED IN THEM.

The pieces should be about three feet long. Then shape them as shown in Fig. 80, the top and bottom pieces being twisted to fit the shaft. Then take a drill brace made as shown in Fig. 81.

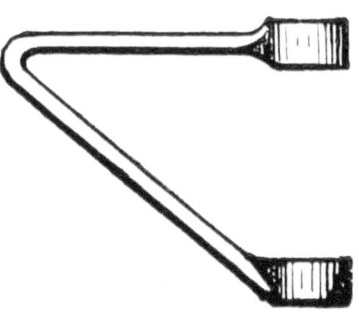

FIG. 80—THE BARS BENT.

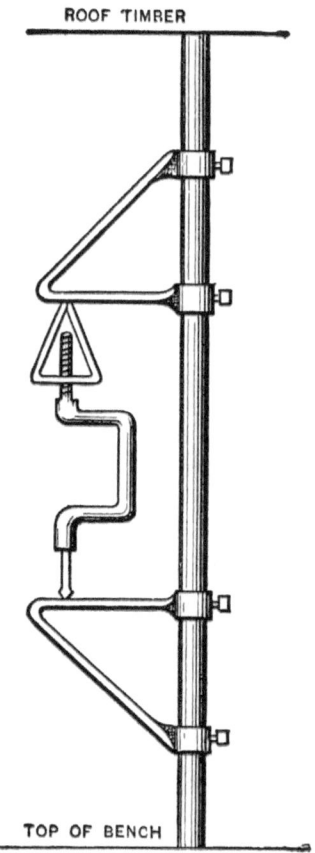

FIG. 81—THE DRILL COMPLETED.

The top part of brace works as a ratchet. The four eyes on the shaft have set screws to hold them in position. The pieces shown in Figs. 80 and 81 can, by means of the set screws, be easily adjusted to the work to be drilled, and will take in larger work than most drills. The drill stock works the same as a ratchet drill. The bottom side of the top cap should be slotted so as to hold the drill stock in place. The thread on the drill can be made four inches long or longer as desired. The threads should be cut very coarse so that they will not wear out too soon. Fig. 81 shows the drill completed.

This drill can be used to good advantage on steam boilers and other machinery that cannot be brought into the shop. —*By* J. W. J.

A HOME-MADE DRILL.

I make a drill brace as follows:

I take a round rod of iron, size 1 ¼ inches, shape it as shown in Fig. 82 of the accompanying illustration, upset it at *A*, and make it about 1 ½ inches square there. I then punch a hole in this end about the same as if for an old-fashioned bit brace, to receive the drill shank.

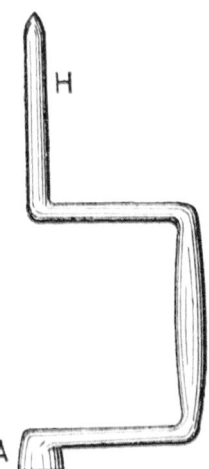

FIG. 82—A HOME-MADE DRILL, SHOWING HOW THE SHANK H IS FORMED AND THE END A UPSET.

I then take a flat piece of iron about a foot long, and draw the ends shown in Fig. 83 at *B, B*. I next take a large nut or a plug of iron 1 ½ inch square, and weld it on the flat piece at *C*, making it 2 inches thick. I then punch or drill a hole large enough to take in the shank *H*, shown in Fig. 84, and cut a very coarse thread to prevent it from wearing loose.

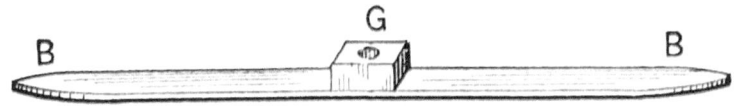

FIG. 83—SHOWING THE FLAT PIECE READY FOR WELDING.

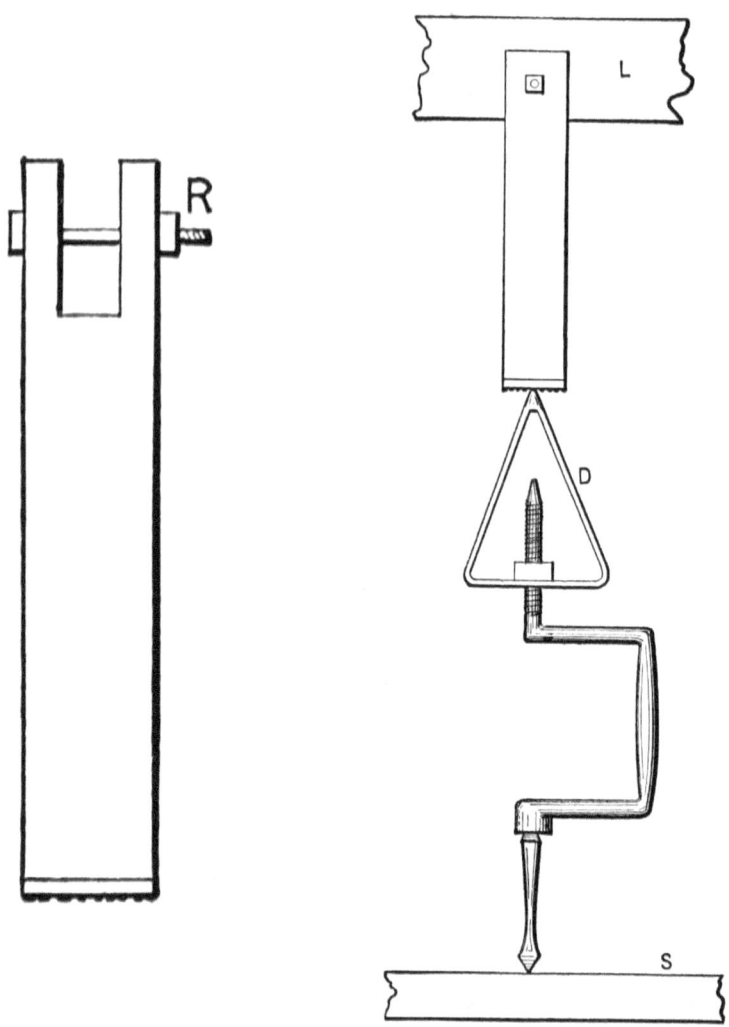

FIG. 84—SHOWING THE PIECE OF SCANTLING USED.

FIG. 85—SHOWING THE JOIST, VISE BENCH, AND DRILL.

I next weld the ends *B, B*, together, and the brace is then ready to be put together as shown in Fig. 85. I next take a piece of scantling 4x4, cut a mortise in one end as shown in Fig. 84, and bore a half-inch hole at *R*, and fasten this scantling over a joist directly over the back of the vise bench, I put a steel plate on the lower end of the piece shown in Fig. 84, and "dot" it well to keep the drill from slipping off. In Fig. 85 the brace is shown completed, *L* being the joist and *S* the vise bench. This is a handy and cheap way to make a drill, and answers well in a small shop where room is scarce.

This make of drill works on the same principle as the ratchet drill, and can be adapted to heavy or light drilling. As the drill cuts itself loose it can be tightened by turning the ratchet D. It is the best home-made drill I know of. —*By* J. W. J.

MAKING AND TEMPERING STONE DRILLS.

My method of sharpening and tempering stone drills may be of interest to some fellow craftsmen. First, in making a drill do not draw down the steel, but cut off each side and then upset back to widen the bit, making strong or light to suit the hardness or softness of the stone to be drilled.

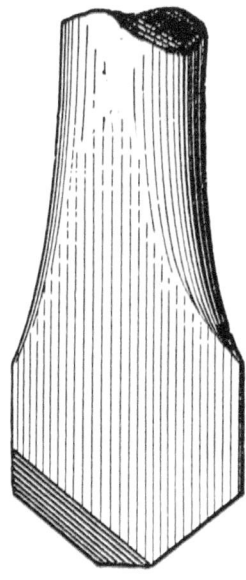

FIG. 86—A STONE DRILL AS MADE BY W. O. WEST.

Next place the drill in the vise and trim off as shown in the accompanying cut, Fig. 86, then lay it down until cool, and then file and temper. Draw the temper twice to a deep blue and you will then have a tool that will drill without cornering a hole, and one that will also stand much better than an ordinary drill.—By W. O. West.

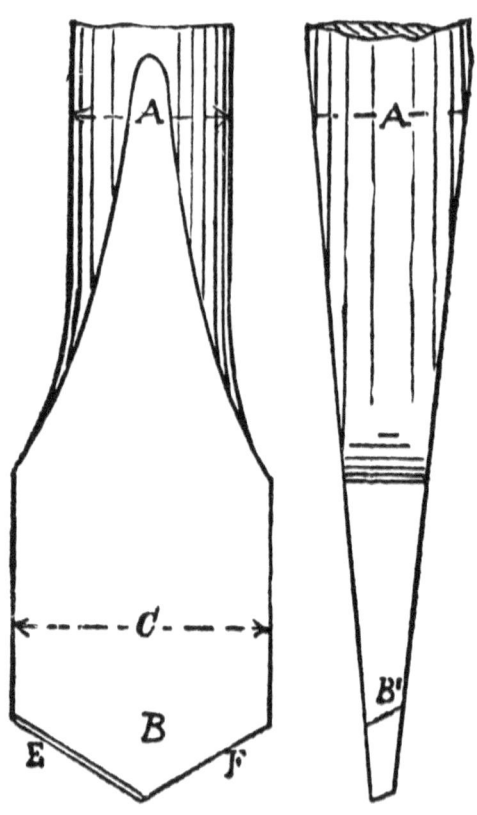

FIG. 87—FORM OF DRILL FOR SMOOTH, STRAIGHT, OR ROUND HOLE.

SOME HINTS ABOUT DRILLS.

To drill a smooth, straight and round hole with a flat drill let the diameter, as at *C*, in Fig. 87, be enough larger than the shank *A* to allow the cuttings to pass freely and parallel, to steady the drill in the hole. Let the bevel at *E* and *F* be, for iron and steel, just enough to clear well, and for crass, give more bevel, as at *B1*. To make a drill cut freely on wrought iron or steel, give it a lip by setting the cutting-edges forward as in Fig. 88.

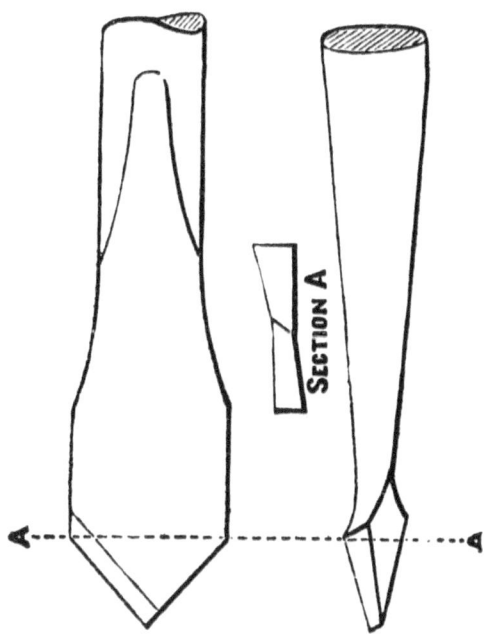

FIG. 88—FORM OF DRILL TO CUT FREELY IN WROUGHT IRON OR STEEL.

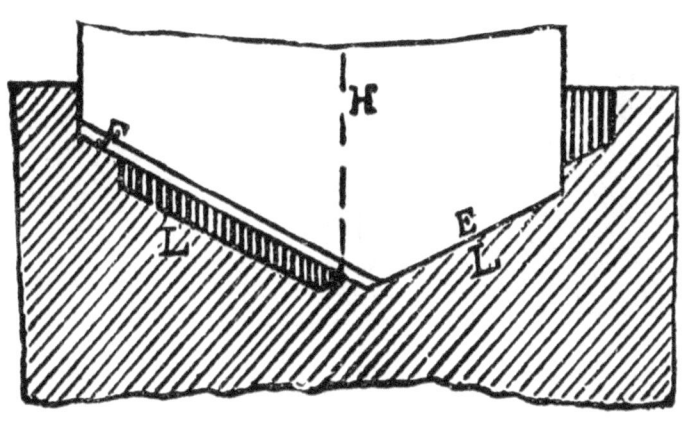

FIG. 89—SHOWING CENTER OF ROTATION AT H.

To make a drill drive easily, first be sure that it runs true and that it is ground true. Suppose, for example that the center of rotation of the drill shown in Fig. 89 is at *H*, and the cutting edges be ground as shown.

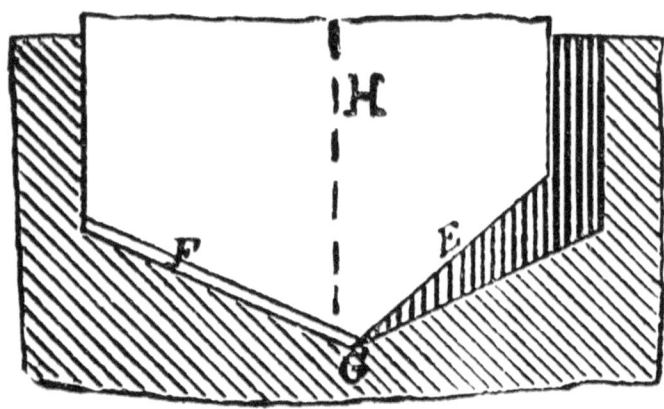

FIG. 90—SHOWS DRILL GROUND TOO MUCH ON ONE SIDE.

Then *E* will cut a certain-sized hole, and *F* will simply act to enlarge it, so that the rate of feed can only be sufficient for one cutting-edge instead of for two. If the drill be ground to one side, as in Fig. 90—*H* being the center of rotation—all the cutting will be done by the edge *F*, and the rate of feed must again be only one-half what it could be if both edges acted as cutting-edges.

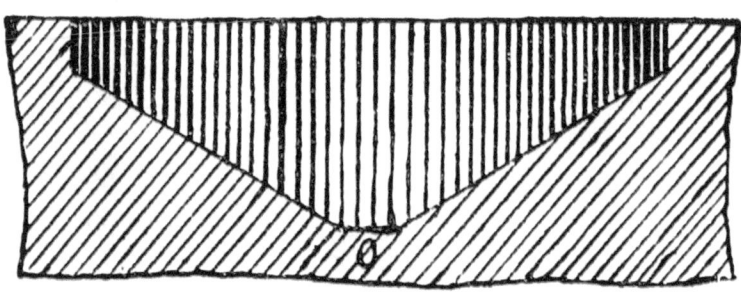

FIG. 91—SHOWING ANOTHER IMPROPER WAY OF GRINDING.

Another secret in making a drill cut easily is to keep the point thin, so that it shall not cut a flat place at the bottom of the cone, as shown in Fig. 91 at *O*, which increases the force necessary to feed the drill.

Drills for brass work should have the cutting edges form a more acute angle one to the other than drills for the fibrous metals, such as steel or iron.

Oil should be supplied to a drill when used on wrought iron, steel, or copper, but the drill should run dry on cast iron, brass and the soft metals, such as babbitt metal, tin, lead, etc.; but very hard steel is easier cut dry than with oil.

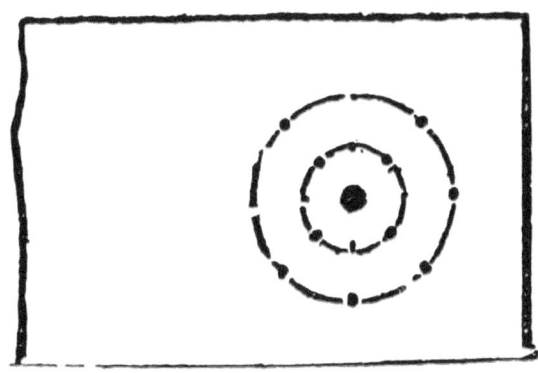

FIG. 92—SHOWING HOW TO MARK A PIECE OF WORK TO BE DRILLED.

For all ordinary work, a drill should be tempered to a bright purple, but for extra hard metal a brown temper may be used.

To drill a hole very true to location, mark it as in Fig. 92, the outer circle being of the diameter of the hole to be drilled, and the smaller one simply serving as a guide to show if the drill is started true. Both circles should be defined by small center-punch marks, as shown, as the oil and cuttings would obscure a simple line.

DRIFTS AND DRIFTINGS.

The drift is a useful tool, once extensively employed, which has been pushed aside by improved machinery. Still, as many a country shop is unsupplied with a slotting machine, the drift may often be used with advantage yet. Indeed no other hand tool will cut with precision a small angular hole where there is no thoroughfare; and even where the tools can pass through, if the metal be thick or there be a number of holes to cut, the drift will be more economical than the file.

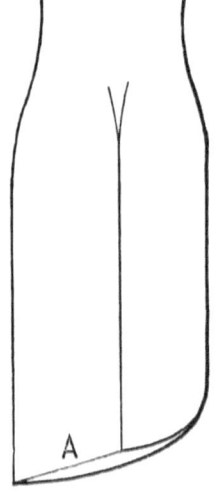 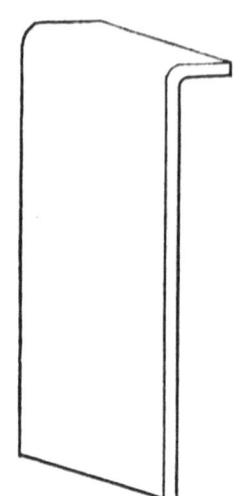

FIG. 93—SHOWING THE SIMPLEST FORM OF DRIFT.

FIG. 94—SHOWING THE STRIP USED IN MAKING A SECOND CUT.

The simplest form of a drift is a steel plug, as shown in Fig. 93 of the accompanying illustrations. This is used to dress out a square hole, especially one with no thoroughfare. It is driven into the hole, the edge A cutting a chip as it descends. The underside slopes back slightly from the cutting face to allow room for chips when the hole does not go through the metal.

To take a second cut, a thin strip of brass or steel, such as shown in Fig, 94, is inserted behind the drift before driving it again, and further cuts are taken by backing up the drift with similar strips until the hole is cut to gauge.

The commonest form of drift is made of a tapered bar of steel, around which teeth—about eight to the inch—are cut with a file, as in Fig. 95. It will readily be seen how a round hole is squared, or a square hole enlarged by driving such a drift through it. The teeth will cut better if filed somewhat diagonally. The tool being very hard, must be fairly struck or it will be liable to break at one of the notches. A round hole can be converted into a hexagonal one in a similar way by means of a six-sided drift, more quickly and much more exactly than by filing.

To square a round hole with no thoroughfare, in a box wrench for instance, it is first made flat on the bottom, say with a D bit.

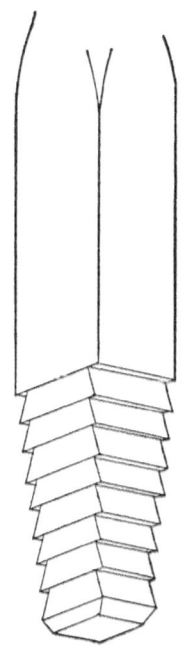 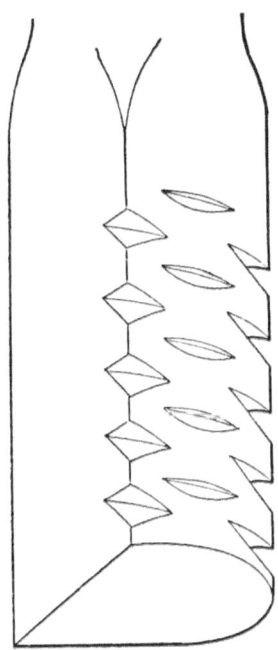

FIG. 95—SHOWING THE COMMONEST FORM OF DRIFT.

FIG. 96—SHOWING A DRIFT USED FOR MAKING ROUND HOLES OVAL.

Then a half round plug is inserted as backing, and a flat drift driven in and gradually fed to the work by strips of brass till half of the hole is cut square. The half round plug is then withdrawn, the drift faced the other way, and the other half of the hole cut, suitable backing being inserted. A half round drift, cutting on its flat face, could be used for this job, instead of the flat one, but would not be so easily backed up and directed.

Half round drifts, cutting on the round face as shown in Fig. 96, are used to make round holes oval, in hammer eyes for instance. They are backed up first by a half round plug and then fed to the work by slips of brass or steel. In this instance, however, the half round plug must have a shoulder, as shown in Fig. 97, to keep it from slipping through the hole.

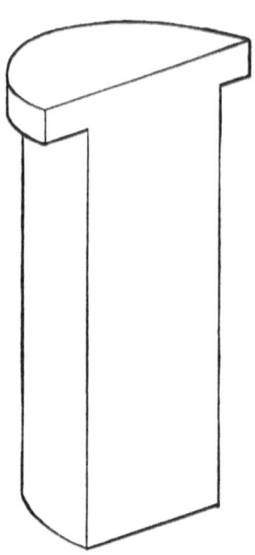

FIG. 97— SHOWING HOW THE SHOULDER IS MADE ON THE PLUG.

These are but the simplest styles of drifting, but they show that the drift can be used to cut almost any shape of hole. —By Will

CHAPTER V.

FULLERING AND SWAGING.

THE PRINCIPLES OF FULLERING.

I should like to say a few words about swaging, which will, I think, be of interest to the younger members of our trade if not to the older ones. Suppose, then, that we take a bar of iron, such as *B*, in Fig. 98, and forge on it a square recess.

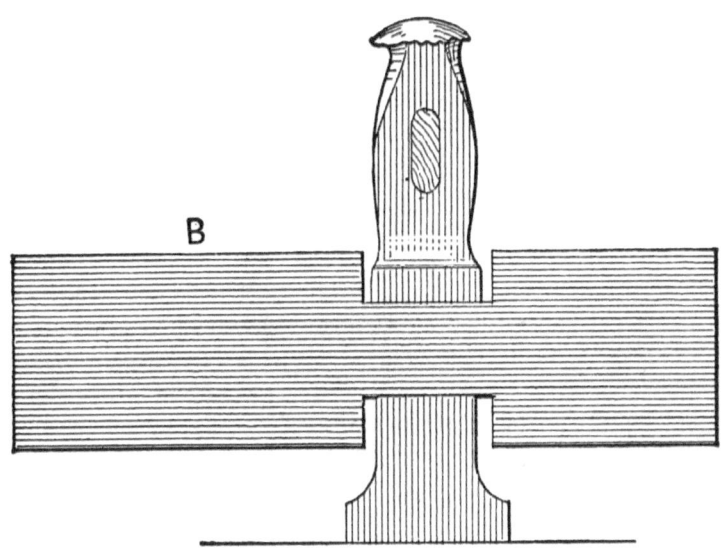

FIG. 98—A BAR IN WHICH A SQUARE RECESS IS BEING FORGED.

The bar will get a certain amount longer, and, in the neighborhood of the recess, a certain amount wider also. Just what the amount would be under any given condition is a matter concerning which I am not aware, but it would, doubtless, vary with the shape and size of the bar, and perhaps also with its degree of temperature I should suppose that the greater the heat the more the metal would spread sideways and the less the bar would elongate, but I may be wrong in this view.

This is a matter of more importance to blacksmiths than at first sight appears. Suppose, for example, a blacksmith is given a pair of dies to make some drop forgings with; in selecting the best size and shape of bar, the question at once arises as to how much the bar will spread in each direction under the action of the blows.

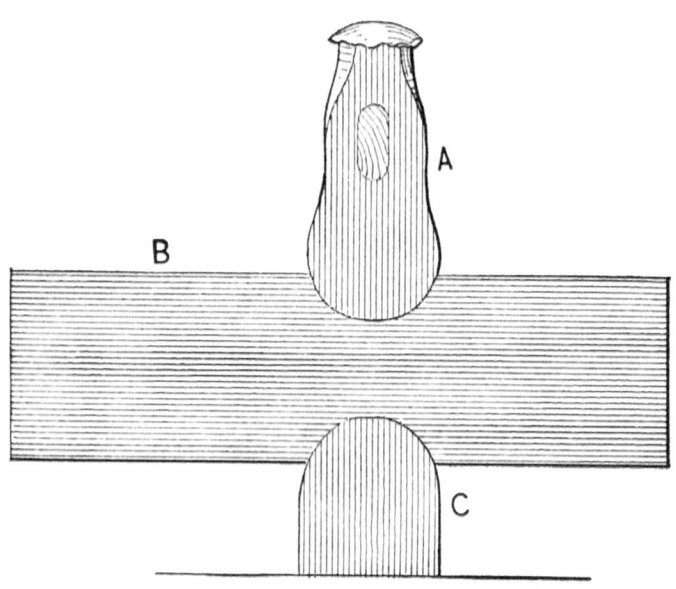

FIG. 99—A ROUND FULLER SUBSTITUTE.

To take a specific case, suppose we require to forge in a die some blocks of iron that must measure 15/16 inch by 1 1/16 inch, and be, say, four inches long, there being enough taper on them to permit their easy extraction from the die; now, what would be the best size of iron to use, and how far should it be placed in the die? Would it be better to cut off the pieces, lay them in the die, and let the blows spread them out, or to take a bar, place it a certain distance over the die and depend upon the elongation of the bar to fill the die?

One could, of course, make an experiment for any given job, but it seems to me there could be got from experiments a rule upon the spreading of iron under compressive blows that would be of great usefulness.

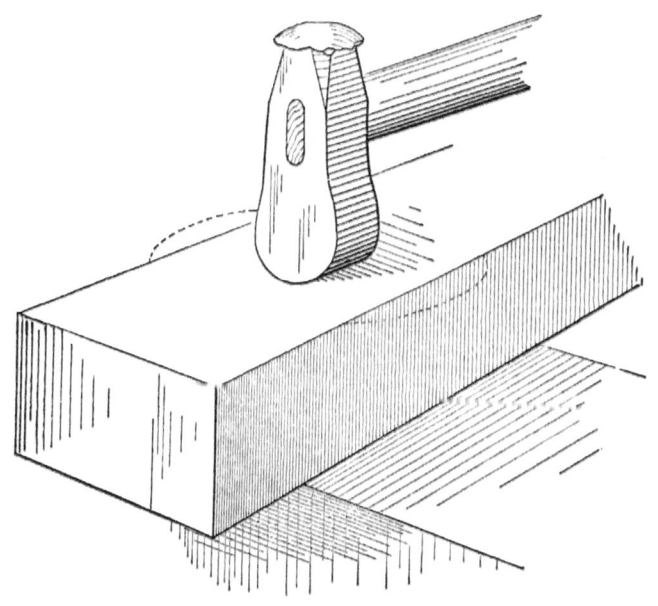

FIG. 100—SPREADING A BAR.

If, instead of a square fuller, we take a round one, as in Fig 99, *A* and *C* representing top and bottom fullers, and *B* the bar, the effect is to increase the elongation of the bar and diminish its spread across the width.

If we require to spread the bar as much as possible and increase its width, we turn the fuller around, as in Fig. 100, causing the spreading to occur, as denoted by the dotted lines.

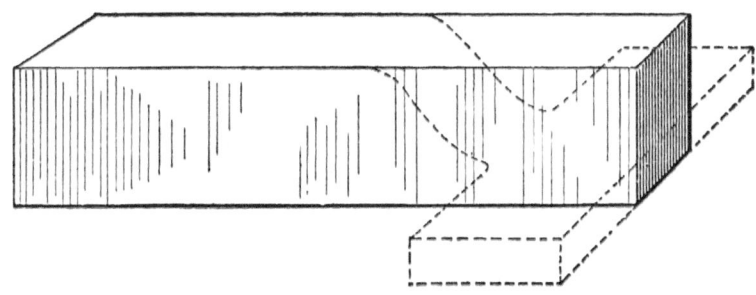

FIG. 101 —THE END OF A BAR TO BE FORGED AS SHOWN BY THE DOTTED LINES.

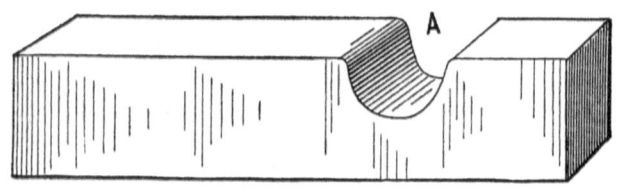

FIG. 102—THE FIRST OPERATION.

Suppose it were required to forge the end of the bar in Fig. 101 to the shape denoted by the dotted lines, the first operation would be to fuller, as at *A*, Fig. 102. Then the fuller would be applied as in Fig. 103, being slanted, as shown, to drive the metal outwards, as denoted by the arrows.

Thus the fuller is shown to require considerable judgment in its use, and to be one of the most useful blacksmith tools.

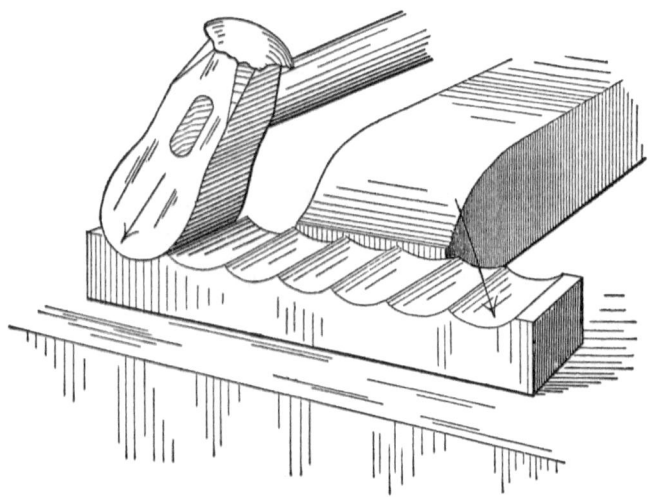

FIG. 103—THE FULLER APPLIED TO DRIVE THE METAL OUTWARD.

If we were to follow the plan of the scientific men, we could very easily claim that a flatter is simply a fuller, on the ground (as the scientists state with reference to gear wheels and racks) that if we suppose the radius of a fuller curve to be infinite in length, then a portion of its circumference may be represented by a straight line; hence, a flatter becomes a fuller whose radius is of infinite length. —*By* R. J.

ABOUT SWAGES.

The old method of forging a swage was to take a piece of the best quality band iron and roll it up to make the body of the swage, and then weld a face of shear or double shear steel on it, the finished tool appearing as in Fig. 103 ½. This forms a good and durable tool, possessing two advantages: First, that a chisel-rod can be used as a handle, instead of requiring to have an eye punched and a trimmed or turned handle; and secondly, that when the head is worn down a new one can readily be welded on, and the tool need not be thrown away.

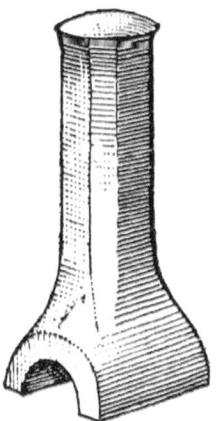

FIG. 103½—OLD METHOD OF FORGING A SWAGE.

In modern practice, however, solid steel swages are employed which, on account of the cheapness of steel and the cheapness of production when made as at present in quantities, obviates to a great extent the necessity for a blacksmith to forge his own tools.

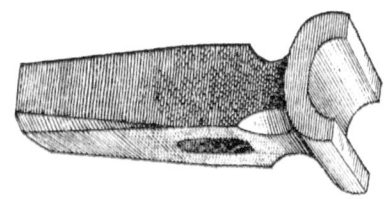

FIG. 104—SEMI-CIRCULAR TOP SWAGE FOR ROUND WORK,

A swage for round work may be semicircular, as in Fig. 104, which represents a top swage only, or V-shaped, as in Fig. 105, which shows a top and bottom swage. There is, however, this difference between the two. That shown in Fig. 104 makes a neater and more truly circular job, but is more apt to draw the work hollow than that shown in Fig. 105.

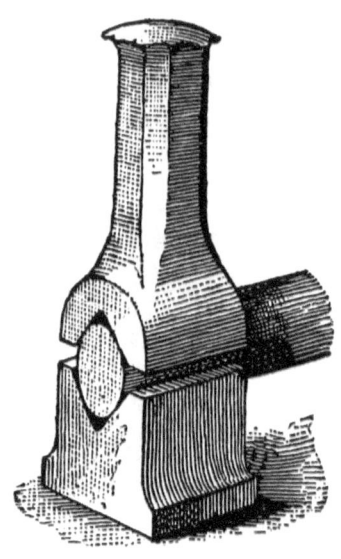

FIG. 105—V-SHAPED TOP AND BOTTOM SWAGE.

The reason of this is that the impact of Fig. 104 is on two sides only, tending to crush the work to an oval instead of closing it to the center, while that shown in Fig. 105 compresses the work on four sides, and prevents its bulging sideways. When iron is compressed on both sides, and liberty is given it to move sideways, the fibers are apt to work one past the other, and a sort of disintegrating process sets in, so that, if the forging be carried to excess on these two sides, without being carried on on the two sides at right angles, the iron will finally split. But if the compression is carried on on four equidistant sides, the forging may be carried to an indefinite extent without separating the fibers, or "hammering hollow", as it is termed. For these reasons, the form shown in Fig. 104 is used to finish work, and, indeed, is used exclusively on small work; while on large work, under steam-hammers, the form shown in Fig. 105 is used to rough out the work, and that in Fig. 104 to finish with. It is obvious that heavier blows may be used without injury to the iron, while the form shown in Fig. 105 is used, the shape of the working face only being, of course, referred to.

FIG. 106—SHOWS A SPRING SWAGE FOR LIGHT WORK.

On very light work, when the hand-hammer only is used, a spring swage, such as shown in Fig. 106, is often used because the top swage guides itself, and the operator, holding the work in one hand and the hand-hammer in the other, is enabled to use the swage without the aid of a helper.

Another method of guiding a small swage is shown in Fig. 107, in which the bottom swage is shown to contain a recess to guide the top one by inclosing its outside surfaces.

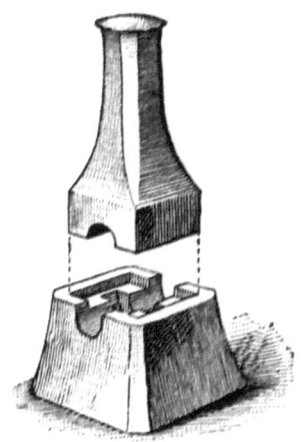

FIG. 107—RECESSED SWAGE.

The holes of circular or round swages are always made of larger circle than the diameter of the work, so that the hole between the two swages, when placed together, will be an oval. This is necessary to prevent the swage hollow from wedging upon the work, and it becomes obvious that in consequence of this form the hollow of the swage must not envelop half the diameter. In practice it usually envelops one-third, or, in large work, still less.

In collar-swages, such as shown in Figs. 106 and 107, the recess for the collar is (to prevent the work from wedging in the recess) made narrower at the bottom than at the top, so that the work may easily be revolved by hand and easily removed from the swage.

Swage-blocks, such as shown in Fig. 108, should have the holes passing through them, as at A, a true circle or square, as the case may be, and parallel for the full length of the hole. But the recesses, B, should be oval, as in the case of hand-swages. Swage-slots, such as shown at C, should, for parallel work, be parallel in their lengths, but taper in their depths, being narrowest at the bottoms of the recesses or slots.

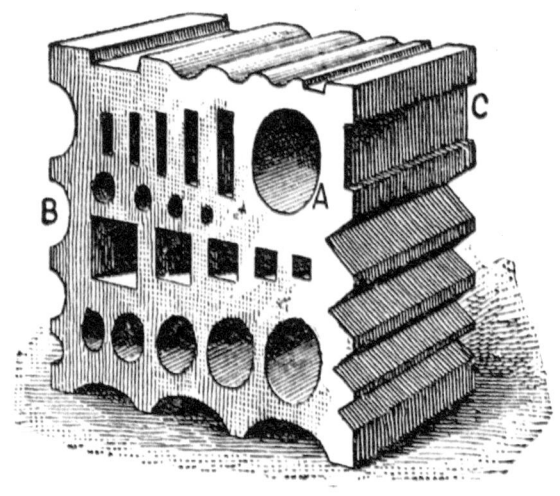

FIG. 108—SWAGE BLOCK FOR GENERAL WORK.

Swages for steam-hammers should have flanges on two opposite sides, as shown in Fig. 109, in which B is an anvil block, S a block swage, and S a hand-swage, and this flange should pass below and envelop the angle block so as to prevent the swage from moving the anvil-block when the work is pushed through the swage; and it follows that the flanges should be on the sides of the swage where the swage-hole emerges, or in other words, the length of the flanges should be at a right angle to the working curved face of the swage. The handle, H, of the hand-swage should be below the level of the striking face, S, so that the hammer-face shall, in no case strike it, which would cause it to vibrate and sting, or perhaps injure the operator's hands.

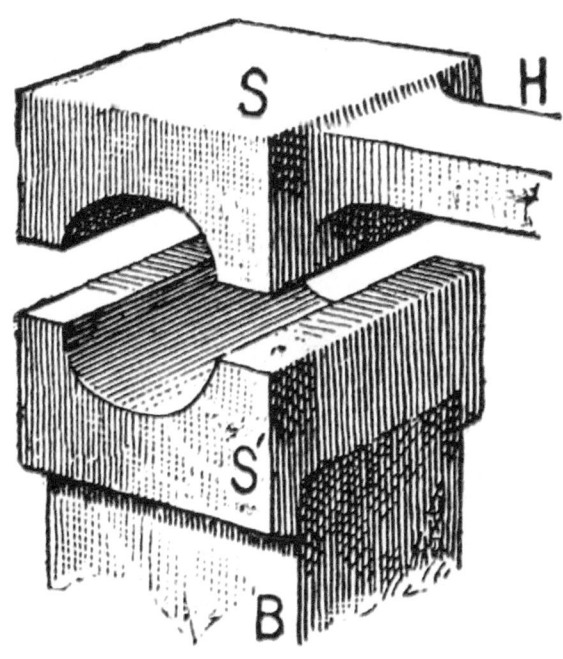

FIG. 109 — FORM OF SWAGE FOR STEAM HAMMERS.

If the position of the steam hammer or other causes renders it necessary, on account of the length of the work, to have the length of the swage-block run at a right angle to the hammer (as in the case of long work done under a trip-hammer, where the parts of the machine under the helm would be in the way of the work), then the flange must fit the sides instead of the front of the anvil-block, as shown in Fig. 110. For small work intended to be neatly finished under a trip-hammer, hinged stamps, finishing tools or dies are often used. Suppose, for example, it be required to forge pieces such as shown in Fig.

111; then, after being roughed out, they may be neatly and cleanly finished to size and shape in a pair of hinged dies, such as shown in Fig. 112.

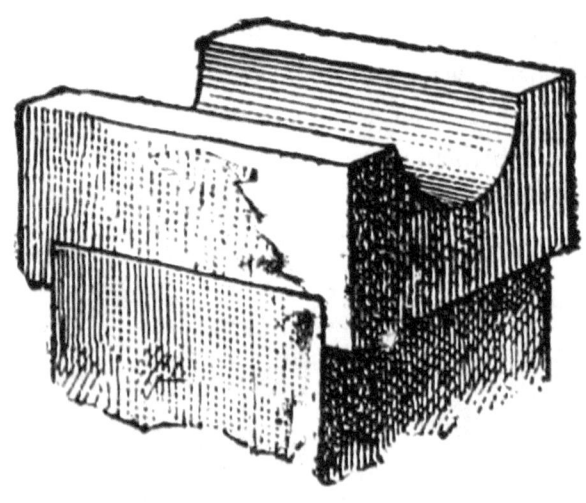

FIG. 110—ANOTHER FORM OF SWAGE FOR STEAM HAMMERS.

The curves in the dies, however, require to be of larger radius than those of the work, so that they may not jam the work and prevent it being revolved in the dies. But the depth of the recess in the dies is made correct for the diameter of the dies, so that when the faces of the two halves of the die meet the work will be of correct diameter.

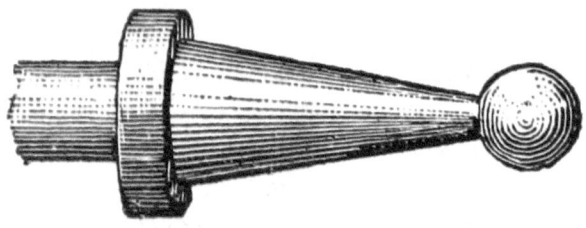

FIG. 111—SPECIMEN OF FORGING.

To free the dies of the oxidized scale falling from the forging, a supply of water is necessary, otherwise the scale would drive into the work-surface, making it both rough and hard to cut. Sometimes, instead of the pivoted joint,

P, Fig. 112, the ends are composed of a spring similar to that shown in Fig. 106, which enables the flat faces of the dies to approach each other more nearly parallel one to the other.

FIG. 112—HINGED DIES FOR FORGING FIG. 111.

RULES FOR SWAGING.

To make a good jump weld it requires good judgment on the part of the smith in getting the two pieces fullered properly before welding. Many smiths do not think or use good judgment when making a weld of this kind.

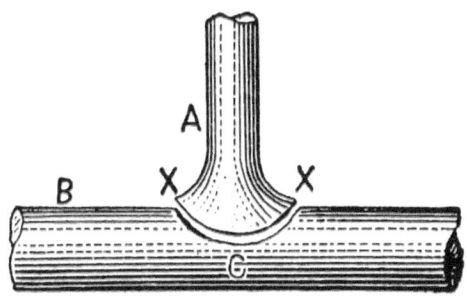

FIG. 113—CORRECT METHOD OF FULLERING OR SWAGING.

To make a jump weld for a shank, carriage-step, or for any other purpose, proceed as in Fig. 113. Fuller at C, upset the shank A, as at the projecting parts, X, X.

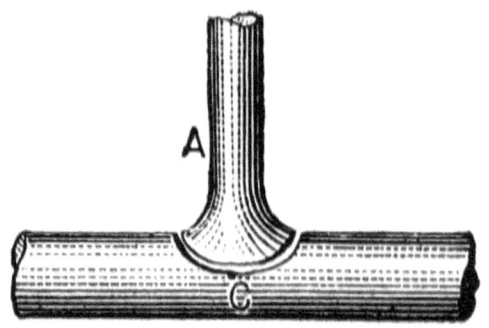

FIG. 114—INCORRECT METHOD.

The shank, *A*, *at X, X*, is the important point to take notice of when making the weld. Always let the shank, *A*, extend over the fullered part, *C*, as at *X, X*. This will give you a good chance in using the fuller when welding so as to get the scarfs, *X, X*, solid to the part *B*.

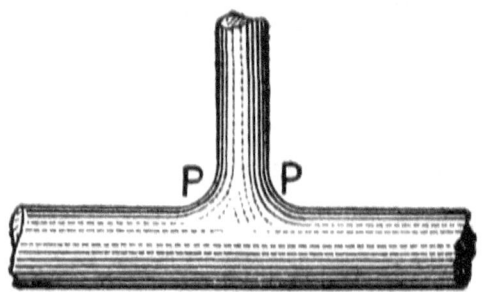

FIG. 115—CORRECT WAY OF FINISHING.

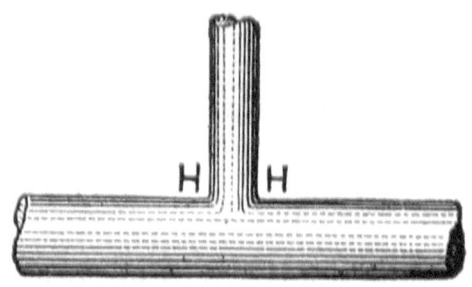

FIG. 116—INCORRECT WAY OF FINISHING.

Never fuller the part *C*, or forge the part *A*, as is shown in Fig. 114. If you do, you will not get a solid weld. To make a neat as well as a strong job, finish as is shown at *P, P*, in Fig. 115. Never finish as is shown at *H, H*, in Fig. 116. A weld made as at *H, H*, in Fig. 116, is not as strong as if made as shown in Fig. 115. —*By* Now and Then.

A STAND FOR A SWAGE-BLOCK.

A blacksmith of my acquaintance once abused the swage-block because he stubbed his toes against it.

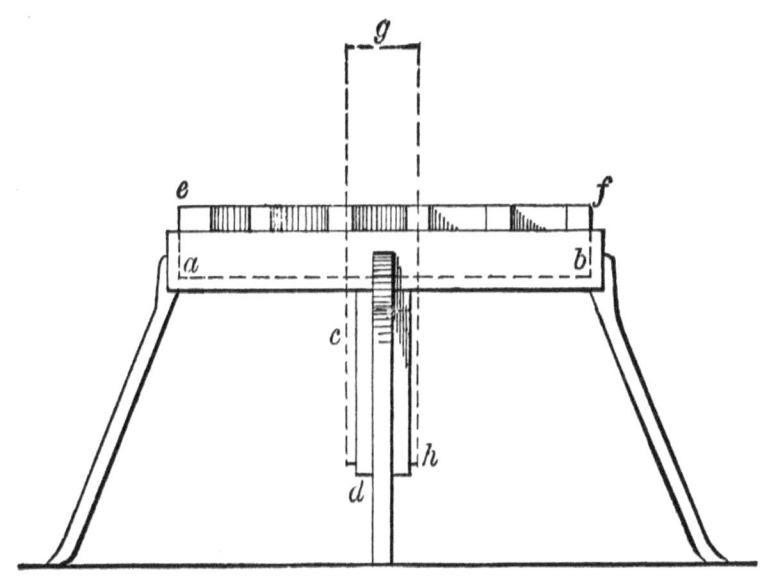

FIG. 117 —STAND FOR A SWAGE-BLOCK.

I want to tell him and others how to save their toes and the swage-block, too.

Let him make a stand for it with four legs, like Figs. 117 and 118, shown herewith. Fig. 117 shows the block, *e, f* lying flat, resting on the ledge shown by the dotted line, *a, b*. The dotted lines, *g, h*, show how the block would stand when upright in the stirrup, *c, d*. Fig. 118 shows the side of the block when upright. —*By* Will Tod.

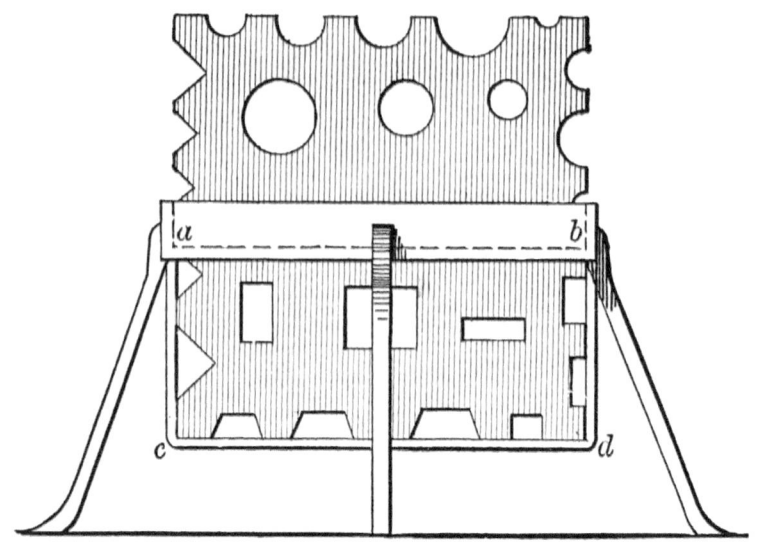

FIG. 118—SHOWING THE SIDE OF THE BLOCK WHEN UPRIGHT.

CHAPTER VI.

MISCELLANEOUS TOOLS.

THE PRINCIPLES ON WHICH EDGE TOOLS OPERATE.

All cutting and piercing edge-tools operate on the principle of the wedge. A brad-awl furnishes an example which all can readily understand. The cutting edge of the awl severs the fibres of wood as the instrument enters, and the particles are compressed into a smaller compass, in the same manner as when a piece of wood is separated by a wedge. A chisel is a wedge in one sense; and an ax, drawing knife, or jack-knife is also a wedge. When a keen-edged razor is made to clip a hair or to remove a man's beard, it operates on the principle of the wedge.

Every intelligent mechanic understands that when a wedge is dressed out smoothly, it may be driven in with much less force than if its surface were left jagged and rough. The same idea holds good with respect to edge-tools. If the cutting edge be ground and whet to as fine an edge as may be practicable with a fine-gritted whet-stone, and if the surface back of the cutting edge be ground smooth and true, and polished neatly, so that one can discern the color of his eyes by means of the polished surface, the tool will enter whatever is to be cut by the application of much less force than if the surfaces were left as rough as they usually are when the tool leaves the grindstone. All edge-tools, such as axes, chisels and planes, that are operated with a *crushing* instead of a *drawing* stroke, should be polished neatly clear to the cutting edge, to facilitate their entrance into the substance to be cut.

HINTS ON THE CARE OF TOOLS.

The following hints on the best means of keeping tools in good condition cannot fail to be useful:

Wooden Parts. —The wooden parts of tools, such as the stocks of planes and handles of chisels, are often made to have a nice appearance by French polishing; but this adds nothing to their durability. A much better plan is to let them soak in linseed oil for a week, and rub them with a cloth for a few minutes every day for a week or two. This produces a beautiful surface, and at the same time exerts a solidifying and preservative action on the wood.

Iron Parts. —*Rust preventives.* —The following receipts are recommended for preventing rust on iron and steel surfaces:

1. Caoutchouc oil is said to have proved efficient in preventing rust, and to have been adopted by the German army. It only requires to be spread with a piece of flannel in a very thin layer over the metallic surface, and allowed to dry up. Such a coating will afford security against all atmospheric influences, and will not show any cracks under the microscope after a year's standing. To remove it, the article has simply to be treated with caoutchouc oil again, and washed after 12 to 24 hours.

2. A solution of india rubber in benzine has been used for years as a coating for steel, iron, and lead, and has been found a simple means of keeping them from oxidizing. It can be easily applied with a brush, and is as easily rubbed off. It should be made about the consistency of cream.

3. All steel articles can be perfectly preserved from rust by putting a lump of freshly-burnt lime in the drawer or case in which they are kept. If the things are to be moved (as a gun in its case, for instance), put the lime in a muslin bag. This is especially valuable for specimens of iron when fractured, for in a moderately dry place the lime will not want any renewing for many years, as it is capable of absorbing a large quantity of moisture. Articles in use should be placed in a box nearly filled with thoroughly pulverized slaked lime. Before using them, rub well with a woolen cloth.

4. The following mixture forms an excellent brown coating for protecting iron and steel from rust: Dissolve 2 parts crystallized iron chloride, 2 antimony chloride, and 1 tannin, in water, and apply with sponge or rag, and let dry. Then another coat of the paint is applied, and again another, if necessary, until the color becomes as dark as desired. When dry it is washed with water, allowed to dry again, and the surface polished with boiled linseed oil. The antimony chloride must be as nearly neutral as possible.

5. To keep tools from rusting, take ½ oz. camphor, dissolve in 1 lb. melted lard; take off the scum and mix in as much fine black lead (graphite) as will give it an iron color. Clean the tools, and smear with the mixture. After 24 hours, rub clean with a soft linen cloth. The tools will keep clean for months under ordinary circumstances.

6. Put 1 quart fresh slaked lime, ½ lb. washing soda, ½ lb. soft soap in a bucket; add sufficient water to cover the articles; put in the tools as soon as possible after use, and wipe them up next morning, or let them remain until wanted.

7. Soft soap, with half its weight of pearlash; one ounce of mixture in about 1 gallon boiling water. This is in every-day use in most engineers' shops in the drip-cans used for turning long articles bright in wrought iron and steel. The work, though constantly moist, does not rust, and bright nuts are immersed in it for days till wanted, and retain their polish.

NAMES OF TOOLS AND THEIR PRONUNCIATION.

Pane, Pene, Peen, which is correct? Pane is the correct word for the small end of a hammer head, Pene or Peen being corruptions. As soon as you leave without any necessity or reason the correct word Pane, you enter a discussion as to whether Pene or Peen shall be substituted, with some advocates and custom in favor of both.

If custom is to decide the matter, Pane will have it all its own way, because, of the English speaking people of the earth, there are, say, thirty-six millions in England, four millions in the West Indies, six or seven millions in Australia with the Cape of Good Hope and other English colonies to count in, who all use the original and correct word Pane, besides Canada and the United States; the former having a majority in favor of Pane from their population being largely English, Scotch, etc., and the latter having some of its greatest authorities, Pane-ites and therefore uncorrupted. Don't let us, as Tennyson says,

"Think the rustic cackle of your burg,
The murmur of the world."

Peen may be used in all parts of the country where "Old Fogy" has been, but it is not used where I have been and that is in Great Britain, the West

Indies, South American English-speaking countries, as Guiana, and not in some parts of the United States; or rather by some mechanics in the United States.

The fact is these corruptions are creeping in and creating dire confusion in many cases. For example: A lathe-work carrier or driver has now got to be called a "dog" in the United States. This is wrong, because if the word *carrier* is used as in other English-speaking countries, the thing is distinct, there being no other tool or appliance to a lathe that is called a carrier. But if the word used is "dog" we do not know whether it means a dog to drive work between the centers of the lathe, or a dog to hold the work to a face-plate, the latter being the original and proper "dog."

Again, in all other English-speaking countries, a key that fits on the top and bottom is a "key," while one that fits on the sides is a "feather." Now a good many in the United States are calling the latter a "key," hence, with the abandonment of the word "feather," a man finding in a contract that a piece of work is to be held by a key, don't know whether to let it fit on the top or bottom or on the sides, and it happens that some mechanics won't have a feather when a key can be used, while others won't have a key at any price.

Let us see what has come of adopting other corruptions in the United States, and I ask the reader the following questions: If I ask a boy to fetch me a three-quarter wrench, is he not as much justified in bringing me one to fit a three-quarter inch tap as a three-quarter inch nut wrench? How is he to know whether a solid wrench, hexagon wrench or a square wrench is meant? In other English-speaking countries, an instrument for rotating the heads of tools, and having a square hole to receive such heads, is a wrench. Thus a three-quarter wrench is a wrench that will fit a three-quarter tap. A "wrench" that *spans* the side of a nut, and is open at the end, is termed a "*spanner*." There can be no mistake about it, it is a spanner or a thing that spans.

Now, suppose the "wrench" goes on the end of the nut head, you call it a box wrench, because its hole is enclosed on all sides but one, and it boxes in the bolt head. Thus the term wrench is properly applied to those tools in which the head of the work is enveloped on all sides by the tool (but not of course, at the end or ends).

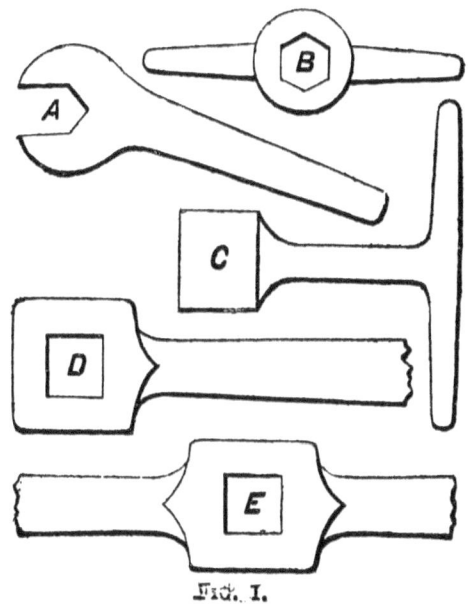

FIG. 119—SHOWS VARIOUS STYLES OF WRENCHES.

For example, in Fig. 119, A is a spanner, B a box wrench, E a double (handled) and D a single wrench, and from these simple elements a name can be given to any form of wrench that will indicate its form and use. Thus, Fig. 120 will be a pin spanner, so that if a boy who did not know the tools was sent to pick out any required tool from its name he would be able to do so if given the simple definitions I have named. These definitions are the old ones more used among the English-speaking people than "monkey wrench," which indicates a cross between a monkey and a wrench.

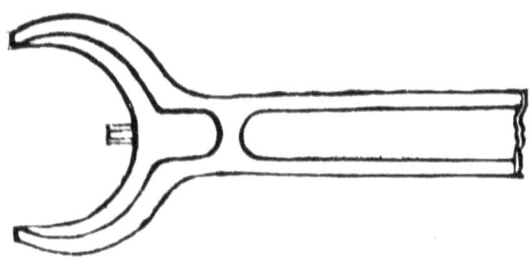

FIG. 120—A PIN SPANNER.

No, no, don't let the errors of a minority influence us simply because we happen to be in a place or town where that minority is prominent, and if we are to make an American language let it be an American improvement, having system and reason in its composition. A man need not say p-e-n-e, *pane*, because the majority of those around his locality were Irishmen and would pronounce it that way whether you spell it pane, pene, or peen. A man need not spell h-a-m-m-e-r and pronounce it *hommer* because the majority of those in the place he is in are Scotchmen. And we need not alter pane to pene or peen promiscuously because a majority of those around us do so, they being in a minority of those speaking our language, especially since pene or peen does not signify the thing named any plainer than pane, which *can* be found in the dictionary, while the former *can't* be found there.

We have got now to some Americanisms in pronunciation that are all wrong, and that some of our school-teachers will insist on, thus: d-a-u-n-t-e-d is pronounced by a majority of Americans somewhat as darnted, instead of more like dawnted: now, if daun, in daunted, is pronounced darn, please pronounce d, a, u in daughter and it becomes "darter." I shrink from making other comparisons as, for example, if au spells ah or ar, pronounce c-a-u-g-h-t.

We are the most correct English-speaking nation in the world, and let us remain so, making our alterations and additions improvements, and not merely meaningless idioms. —*By* Hammer and Tongs.

Note. —This writer talks learnedly, but nevertheless he is condemned by the very authority which he cites (and correctly too) in support of his pronunciation of the word *Pane*. Webster's Unabridged gives the *au* in *daunted* the sound of *a* in *farther* so that the word (our contributor to the contrary notwithstanding), should be pronounced as though spelled *Darnted*. —Ed.

TONGS FOR BOLT-MAKING.

I send a sketch of a pair of tongs suitable for making bolts. The jaws are eight inches and the reins ten inches long. A glance at the engraving, Fig. 121, will show that it is not necessary to open the hands to catch the head on any size of bolt.

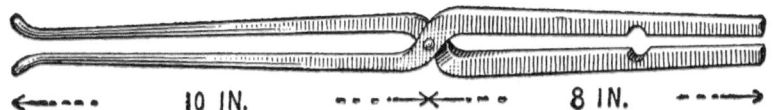

FIG. 121—TONGS DESIGNED BY "SOUTHERN BLACKSMITH."

These tongs should be made very light. The trouble with all nail grabs is that the rivet is put too near the prongs, and when you try to get nails out of the bottom of a keg the reins catch in the top and the tongs can't open far enough. —*By* Southern Blacksmith.

HOME-MADE FAN FOR BLACKSMITH'S USE.

To construct a home-made hand blower proceed as follows: Make two side pieces of suitable boards of the shape shown in Fig 122 of the accompanying sketches. Make a narrow groove in the line marked A. Procure a strip of sheet-iron of the width the blower is desired to be, and bend it to correspond with the groove. Then the two sides are to be clasped upon the sheet-iron, with small bolts.

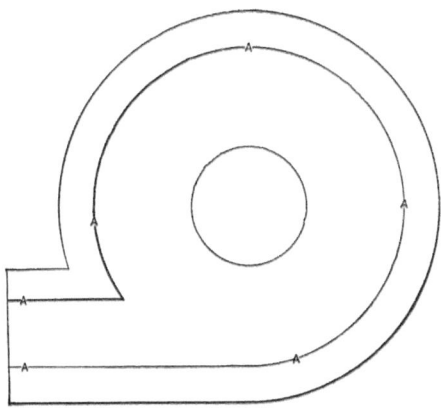

FIG. 122—SIDE ELEVATION OF "E. H. W.'S" BLOWER.

This will form the blower case. The small circle shown in the center of Fig. 122, incloses a portion to be cut out for the admission of the current of air.

The shaft is made by taking a block of wood large enough to make a pulley about 1 ¾ inches in diameter, the length of the block being from bearing to bearing of shaft. Bore a central longitudinal hole ½-inch in diameter in the block, turn a plug to fit the hole. Put the plug in place and place all on the lathe and turn, leaving the part where fans are to be attached about 1 2-5 inches in diameter. Square this part and fasten the fans thereto, as shown in Fig. 123.

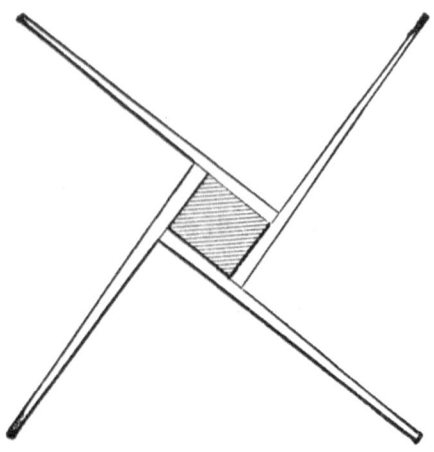

FIG. 123—MANNER OF ATTACHING THE FANS TO THE SHAFT OF BLOWER.

Constructed as there shown they are intended to revolve from left to right. On removing the block from the lathe the wooden plug is withdrawn and a rod of half-inch iron is put in, projecting at each end an inch and a quarter for journals. The bearings for the shaft are simply blocks of wood screwed to the sides of the case, with holes bored to fit the shaft as shown in Fig. 124.

The dimensions of my blower are as follows: Case, 9 inches in diameter inside of sheet-iron; width of case, 3 inches; central opening, to admit air, 3 inches in diameter; pulley, 1 ¾ inches in diameter. The fans are of pine, one-quarter inch thick at the base, diminishing in thickness to one-eighth inch at the point. Fig. 125 shows my portable forge, upon which the above described fan is employed. The frame is made of four-foot pine fence pickets. The fire-pot is an old soap kettle partly filled with ashes to prevent the bottom from getting too hot. —*By* E. H. W.

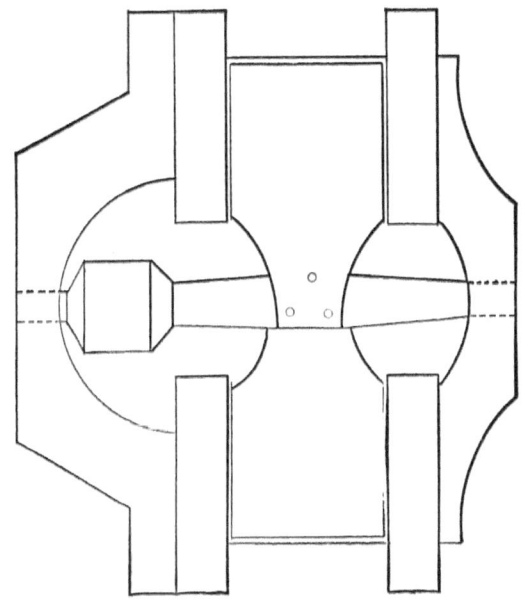

FIG. 124—CROSS-SECTION THROUGH BLOWER, SHOWING BEARINGS FOR SHAFT.

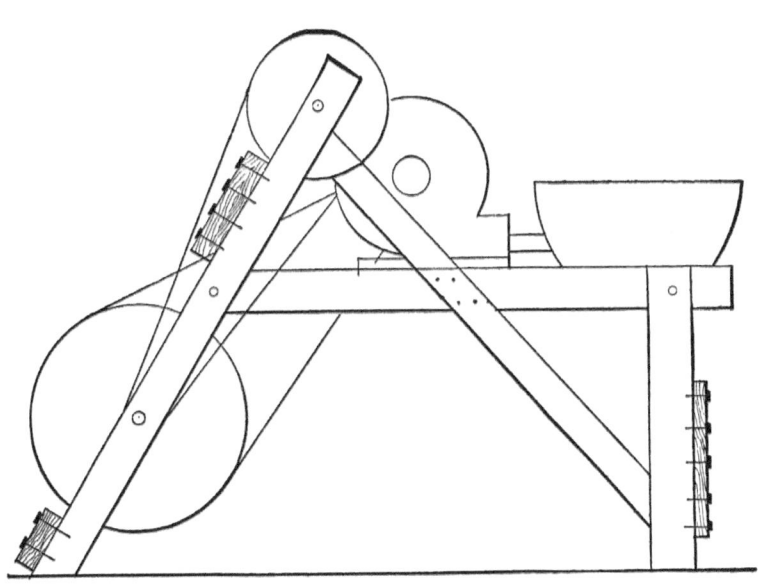

FIG. 125—GENERAL VIEW OF BLOWER, IN CONNECTION WITH FORGE.

MAKING A PAIR OF PINCHERS.

The subject of my remarks is one of the simplest yet most useful tools in the shop, a pair of pinchers. Fully one-half of my brother smiths will say, "Who doesn't know all about pinchers?"

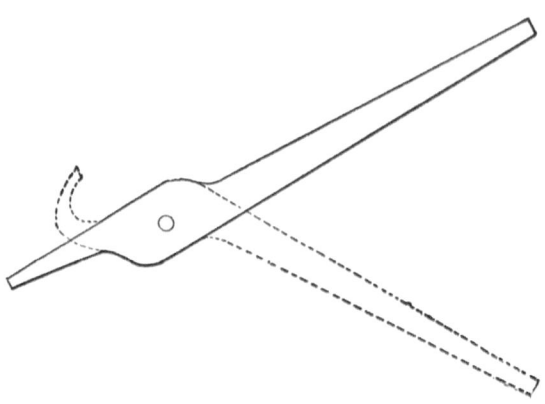

FIG. 126—SHOWS METHOD OF FORGING.

From the appearance of two-thirds of those I see in use, I am convinced that if the makers knew all about them, then they slighted their work when they made them. To make a neat and strong pair of pinchers, forge out of good cast steel such a shape as is shown in Fig. 126, and then bend to shape, as in the dotted lines.

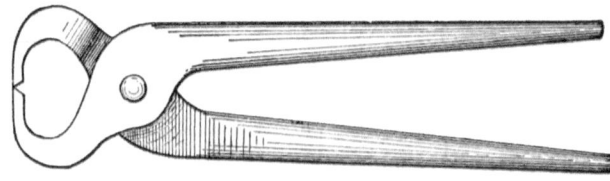

FIG. 127—SHOWING THE PINCHERS FINISHED.

When one-half of the forging is done, forge the other jaw in same way, and make sure that they have plenty of play before riveting them together. Then fit and temper, and you will have a strong neat tool, as shown in Fig. 127. Fig. 128 represents the style of pinchers generally made.

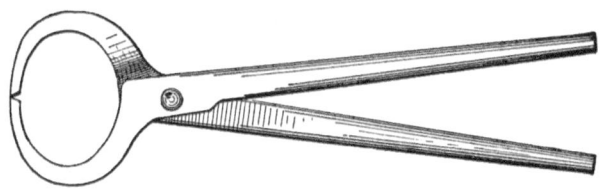

FIG. 128—SHOWING A FAULTY METHOD OF MAKING PINCHERS.

They are awkward and weak looking, and work about as they look. The cutting edge is entirely too far from the rivet. —*By* J. O. H.

A HANDY TOOL FOR HOLDING IRON AND TURNING NUTS.

Fig. 129 represents a small tool I use in my shop and find very handy to hold round iron and turn nuts, etc., in hard places. Wheelwrights and carriage painters, as well as blacksmiths, will find it a very convenient tool.

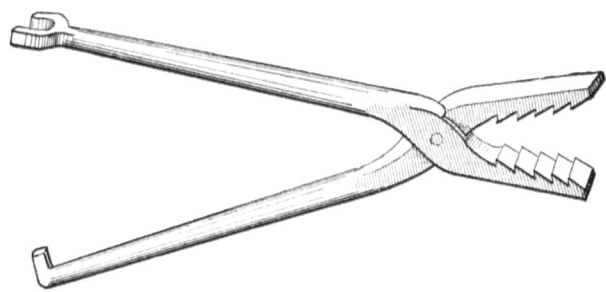

FIG. 129—TOOL FOR HOLDING IRON AND TURNING NUTS.

It is of steel and is quite light. It is made the same as a pair of tongs, having teeth filed on the inside of the jaws and having clasp pullers on the end of the handles. —*By* L. F. F.

A HANDY TOOL TO HOLD COUNTERSUNK BOLTS.

A handy tool to hold countersunk bolts in plows, etc., is made as follows: I take a piece of iron, say ¾-inch square and ten inches long, as shown in Fig.

130, and punch a slotted hole four inches from the pointed end. The hole should be ¾ x ¼ inch. I then take a piece of iron ¾-inch square and make a slotted hole in this at one end and a tenon in the other, as shown at A in Fig. 131. I next punch a ¼-inch hole in the slotted end, and then take a piece of iron ¾ inch x ½ inch and 6 inches long, and draw one end out to ½ inch and turn a hook on it, as shown in Fig. 132.

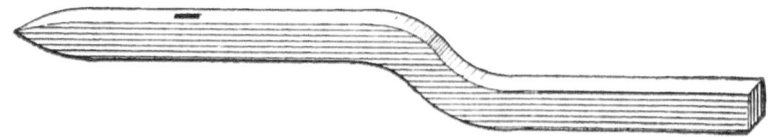

FIG. 130—THE FIRST STAGE IN THE JOB.

I then punch two or three holes in the other end, take a piece of steel ¾ inch x ¼ inch, draw it to a point like a cold chisel; then take the piece shown in Fig. 13, split the end, and put in the small pointed piece I have just mentioned, and weld and temper.

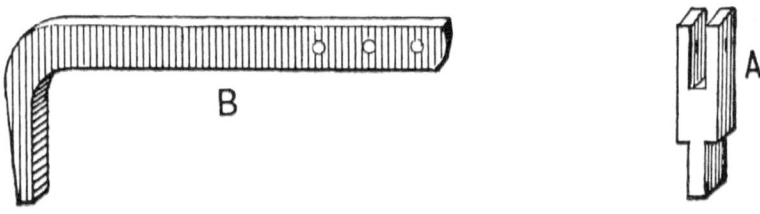

FIG. 131—TWO ADDITIONAL PARTS OF THE TOOL.

I then put all my pieces together in the following manner: After heating the tenon end of the piece *A* in Fig. 131 to a good heat, I place it in my vise and place on the piece shown in Fig. 130, letting the tenon of the piece *A* go through the piece shown in Fig. 130, and while it is hot rivet or head it over snugly and tightly. If this is done right, the tenon and slotted hole in A will point the same way. I then take the piece *B* shown in Fig. 131, and place it in the slot of *A* and join the two pieces with a loose rivet, as shown in Fig. 132, so that the piece can be moved about to suit different kinds of work.

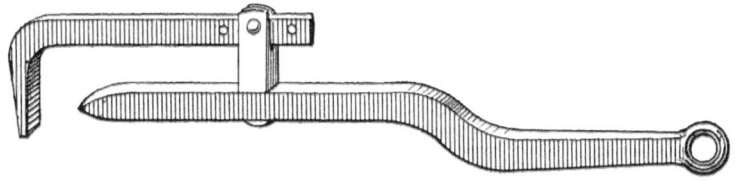

FIG. 132—THE PARTS UNITED AND TOOL COMPLETE.

I next place the pointed end of the bolt holder against the bolt head and give the other end a tap with the hammer; then hook the piece shown in Fig. 131 over the plow bar and bear down on the cutter end with my knee, while with my wrench I take off the top. A ring may be welded in the end to use in hanging the tool up. —*By C. W. C.*

MAKING A PAIR OF CLINCHING TONGS.

The following is my way of making clinching tongs: I take a piece of 5-8 square steel and forge it out the same as I would for common tongs, making the jaws ½ inch wide by 3-8 inch thick, one jaw 1 ½ inches long and the other 2 ½ inches in length. I then draw out the short jaw to 10 inches and draw the long one to 12 inches. I then turn the long jaw back as shown at *A* in Fig. 133 of the accompanying engravings, and shape the short jaw as in Fig. 134. I next take the ¼-inch fuller and notch the inside of the short jaw as shown at B in Fig. 134.

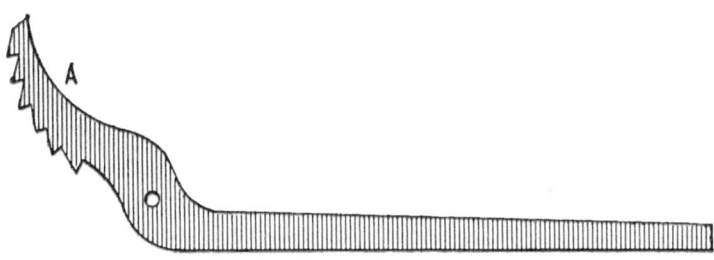

FIG. 133—THE LONG JAW.

I then put notches in the long jaw at *A*, and next drill a 5-16-inch rivet hole as in other tongs and take a 5 16-inch bolt with a long thread and screw one nut on the bolt down far enough to receive both jaws and another nut. I then temper the curved jaw at *A* until a good file can just cut it.

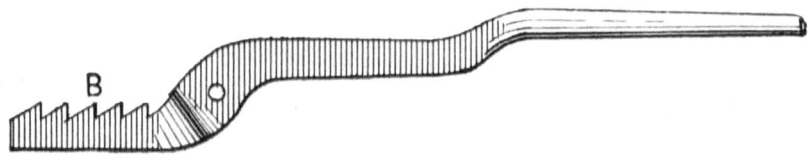

FIG. 134—THE SHORT JAW.

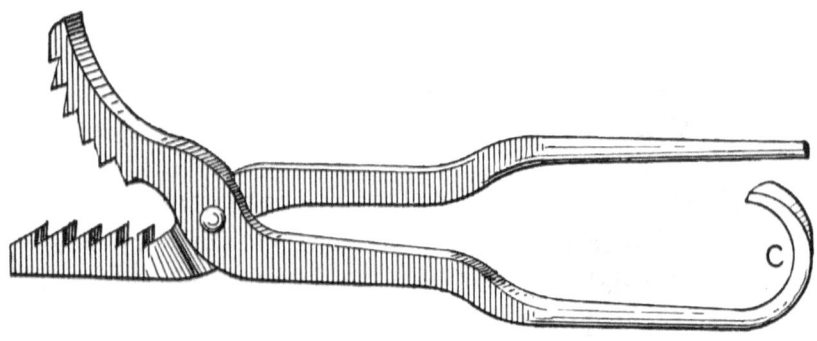

FIG. 135—THE TONGS COMPLETE.

I next put in the bolt and bend the reins as shown in Fig. 135. The object of bending at C is to prevent the jaws from pinching your fingers if they slip off a clinch. I have a pair of tongs made in this way that I have used for the last five years. —*By* J. N. B.

TONGS FOR HOLDING SLIP LAYS.

The accompanying illustration, Fig. 136, represents a pair of tongs for holding right and left hand slip lays while sharpening and pointing, and making new lays. This tool does away with the clamping or riveting of the steel and bar before taking a welding heat, because it holds the two parts together better than any clamp or rivet.

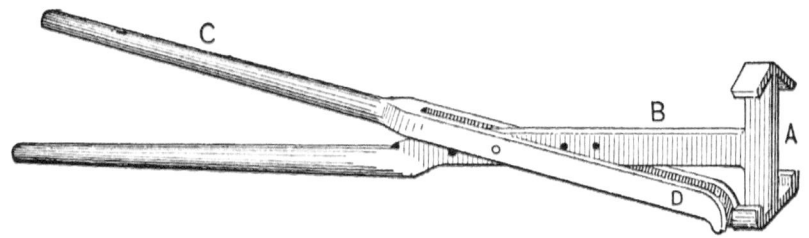

FIG. 136—TONGS FOR HOLDING SLIP LAYS.

The double *T*, or head *A*, is forged from Swedish iron. The handle, *B*, is welded to the head in the center, as shown in the illustration, arid works between the forked handle *C*. *D* pushes against the end of the lay bar (right or left as the case may be), and that draws the top *T* up tight, and as it is bent in the same angle as a plow lay, but little power of grip is needed to hold the tongs to their place while sharpening or pointing.

They clamp up so tight that often you have to tap them on the end of the handles to release the lay.

Both handles have ¼-inch holes through them and I use ¼-inch bolts or pins in them to hold them together after they have been adjusted to fit the lay.

Any good blacksmith can make these tongs and will certainly find them very useful. —*By* A. O. K.

CHAPTER VII.

MISCELLANEOUS TOOLS.
CONTINUED.

MENDING A VISE.

I will make a few remarks about the vise, a tool which blacksmiths use every day. Some smiths believe that when the threads or screws are worn out, it is necessary to buy a new vise. But this is an error; for the old vise can be mended so as to be as good as a new one. The job is done as follows: First cut the old vise screw off where the screw stops, or about two inches from the collar of the large end, then weld on a new piece of round iron and turn threads on it the same as if for a new vise screw. To get them in line these threads should be done with a lathe. It is better to have the screws taper slightly. The next thing in order will be to get the screw threads in the box. First, take a drill and turn the old threads out, smoothly and true. Care must be taken to have a space of at least 3-16 of an inch all round the screw when it is placed in the threadless box. Then take a long piece of Swedish iron just thick enough to be bent in the screw, and about 3-16, or a fraction under, than the depth of the threads on the screws. Bend this flat iron from one end of the screw to the other, then make another piece of iron that will fit in the screws that the first piece of iron left after it was bent. This second piece we find will be as thick as the screws on the vise screw, and as wide as the first piece extends above the screws. This being done, all is closed up smooth, the second piece holding the main threads firm after the screw is turned out. Now the threads are on the vise screw, and it is nearly ready to be put in the box. First dress the outside of the threads off, so the screw can be driven in the box when it is a little warm, then drive screw threads all into the box at about the place where they belong, then let the box cool, and turn the screw out of the threads and braze the screws; put the brass on end side. When you are melting it, keep turning the box so it will be brazed all over. When cool, grease the screw and threads well and slowly turn it in.

The main thread must not fit too tightly in the screws. This makes the best vise screw. By measuring the depth of the hole in the box you can tell how long to make the threads on the screw—*By* J. W.

A CHEAP REAMER.

The following may be an old idea, though I have discovered it for myself: Heat an old three-cornered file; hammer one corner down, then grind the same round and the other corner sharp, temper, and you have a cheap taper reamer, cutting both ways. —*By* Will Todd.

SHAPES OF LATHE TOOLS.

Every toolsmith knows the trouble he has to contend with in tool dressing. One man wants a tool this shape, another a different shape. One wants his tools fully hardened, while another prefers a straw temper. Of course, in a shop large enough to keep a toolsmith for this special work, the smith makes the tools to the shape he has found by experience to be the most suitable for general work, only varying it to suit some special occasion.

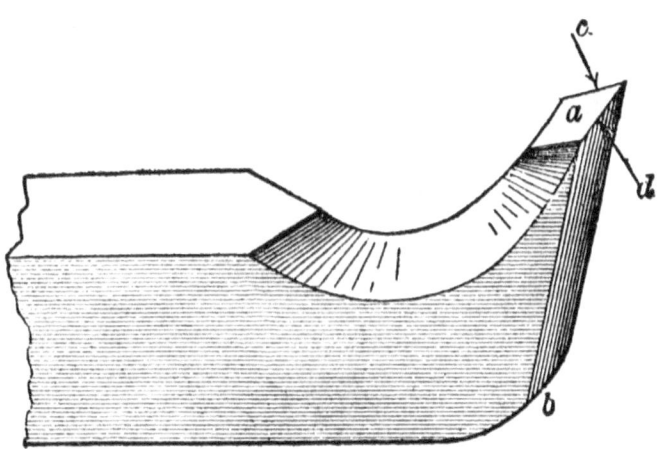

THE SHAPE AND USE OF CUTTING TOOLS FOR LATHES. FIG. 137—
THE DIAMOND POINT TOOL.

But in a small shop tool forging becomes a part of the duty of the ordinary blacksmith, and in order that he may know what shapes tools should be given he must understand the principles governing their action. These principles I will now explain so plainly and carefully that no one can fail to comprehend them.

Fig. 137 of the accompanying illustrations represents what is commonly called the diamond-point tool, a being its top face, b its bottom one, and c and d the cutting edges.

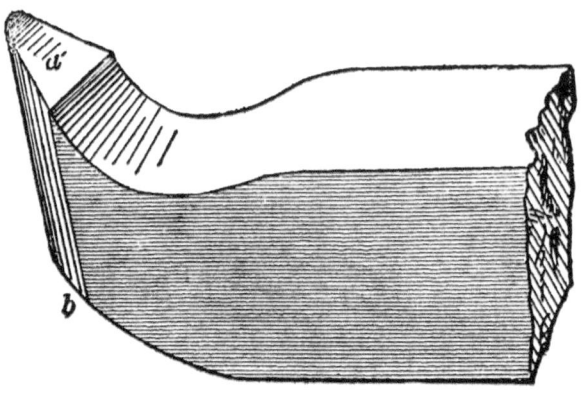

FIG. 138—SHOWING A TOOL EASY TO FORGE AND GRIND.

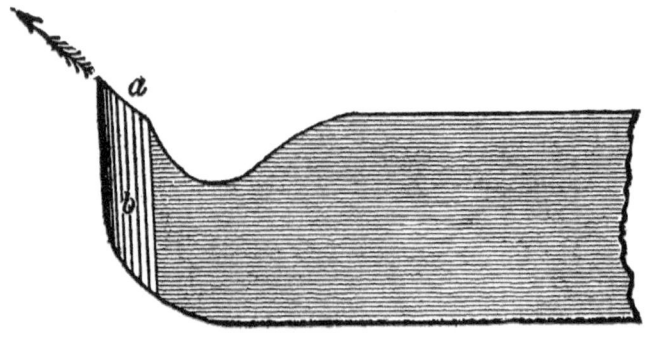

FIG. 139—SHOWING THE RAKE OF THE TOP FACE.

This is a very common form of tool, but I do not believe it is the best one even for the purpose of plain outside turning, a much easier tool to forge and to grind being the one shown in Fig. 138. The rake of the top face is its angle in the direction of the arrow in Fig. 139, and the rake of the bottom face b is its angle in the direction of the arrow in Fig. 140.

The efficiency and the durability of the cutting edge depends upon the degrees of rake given to these two faces. Obviously the less rake the stronger the cutting edge, but the less keen the tool.

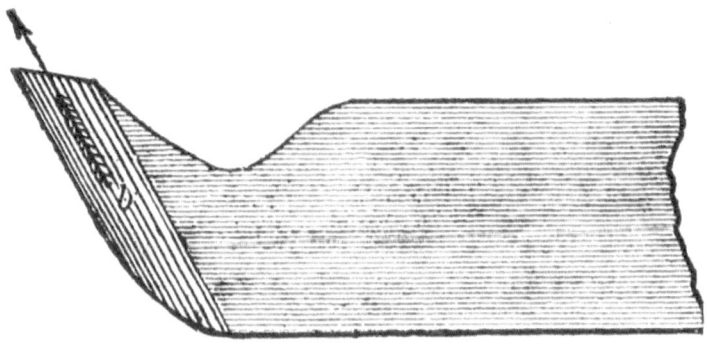

FIG. 140—SHOWING THE RAKE OF THE BOTTOM FACE.

If we give a tool an excess of top rake, as in Fig. 141, the cutting edge will soon dull, but the cutting C will come off clean cut and in a large coil, if the tool is fed into the work.

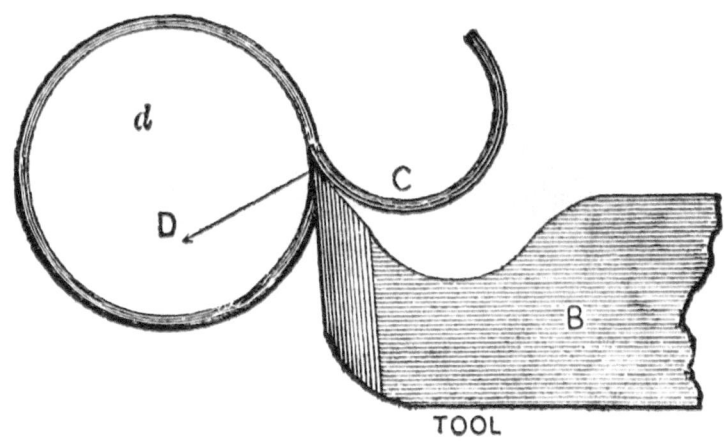

FIG. 141—SHOWING A TOOL WITH AN EXCESS OF TOP RAKE.

The strain on the top face, however, will be in the direction of *D*, and the tool will be liable to dip into the cut when the cut deepens, as it will in some places on account of a want of roundness in the iron.

If, on the other hand, we give too much bottom rake, as in Fig. 142, the cutting edge will be weak, and there being but little top rake, the pressure will be in the direction of *D*. Furthermore, the cutting will come off in almost straight pieces and all broken up.

A fair amount of top and bottom rake for wrought iron is shown in Fig. 143, the top rake being diminished for cast iron and for steel.

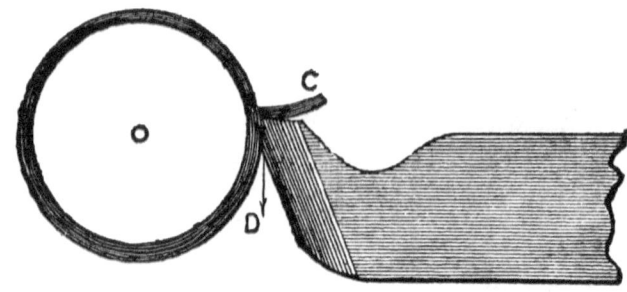

FIG. 142—SHOWING A TOOL WITH TOO MUCH BOTTOM RAKE.

The best guide as to the efficiency of a tool is its cuttings or chips, as they are commonly called.

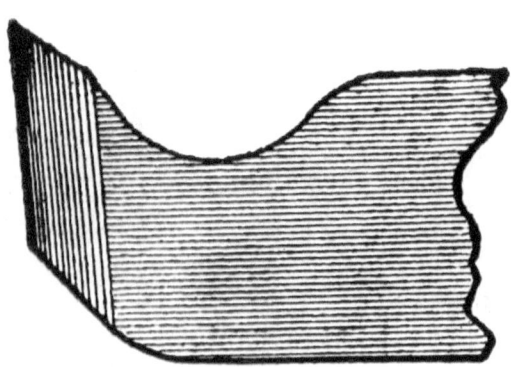

FIG. 143—SHOWING A WELL PROPORTIONED TOOL FOR WROUGHT IRON.

When the tool is fed upward only, the cutting should come off in a large circle as in Fig. 141, and if the coil is small there is insufficient top rake. But when fed level along the work the cutting comes off in a spiral, such as in Fig. 144. The more top rake the tool has the more open the coils will be, the cutting shown being as open as it should be even for the softest of wrought iron. The harder the metal the less top rake a tool should have, while the bottom rake should in all cases be kept as small as possible say 10 degrees for the rate of feed of an average 16-inch lathe.

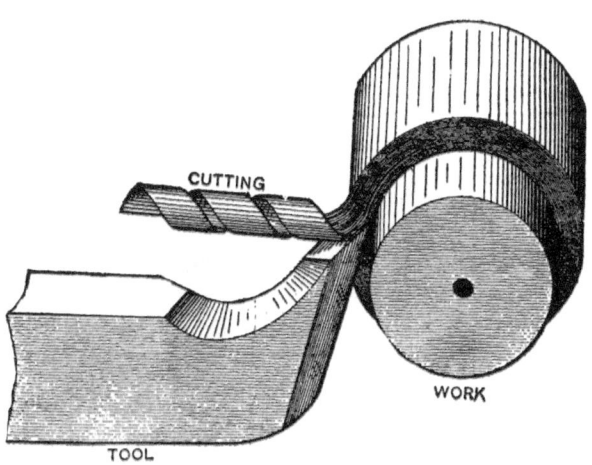

FIG. 144—SHOWING HOW THE CUTTING COMES OFF WHEN THE TOOL IS FED LEVEL.

The tool point should never stand far above the top of the tool steel, the position shown in Fig. 138 being for a newly-forged tool.

The practiced hand may have these tools fully hardened, but the beginner should temper them to a light straw color.

The top of anvil should be perfectly level and smooth, and the center of the same should be harder than the edges, because it receives the brunt of the blows. When used on the upper floors of a building, when a solid foundation can not be had, resort is had to a device by means of which the jar of the blows is obviated. This consists in mounting the anvil upon a stout spring whose upward rebound is counteracted by smaller springs placed above. —*By* Joshua Rose.

USEFUL ATTACHMENT TO SCREW STOCK DIES.

Having an order to fill for a small quantity of iron pins 2 ½ by 1-8 inch in diameter, to be threaded at each end with a wood screw thread, and having no tools to cut the threads with, I devised and used the following plan, which answered so satisfactorily that I think the idea may be of service to others, and hence send you sketches of it.

C, in Fig. 145, is the die stock having in it the dies *B B*, with the requisite pitch of thread. On the die stock is fastened a tool post and tool at F and a copper steadying piece at *A*. The tool may be made from ¼-inch square Stubbs or Crescent steel carefully filed to the shape of the thread to be cut and carefully tempered. *E* is a steel sleeve screwed with a thread of the same pitch and sawn through its axis one way to point *D*, and the spring tempered in oil.

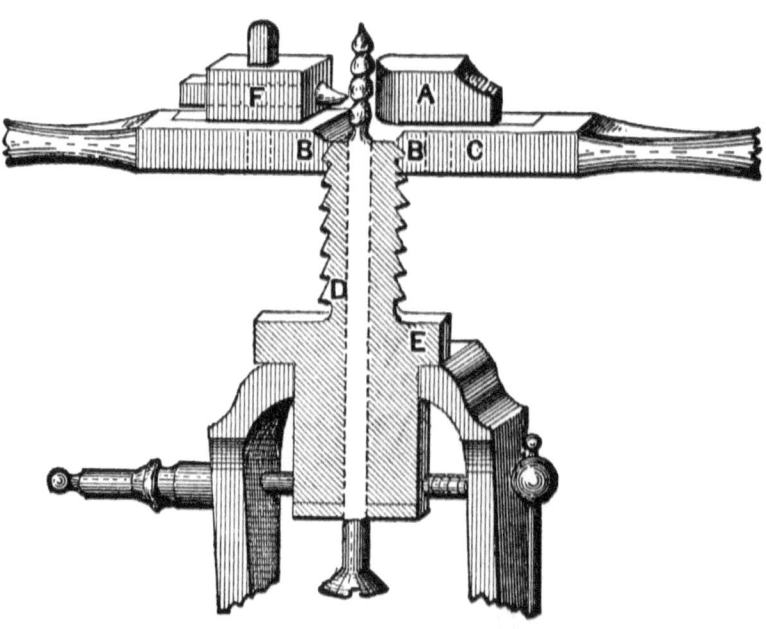

FIG. 145—ATTACHMENT TO SCREW STOCK DIES AS DESIGNED BY "W. D." SIDE VIEW.

The pin to be threaded is inserted through the sleeve *E*, which on being gripped in the vise securely holds the pin. When the stock is revolved E regulates the pitch of the thread that the tool will cut on the pin. Hence the stock may be used in the ordinary manner. A plan view is shown in Fig. 146.

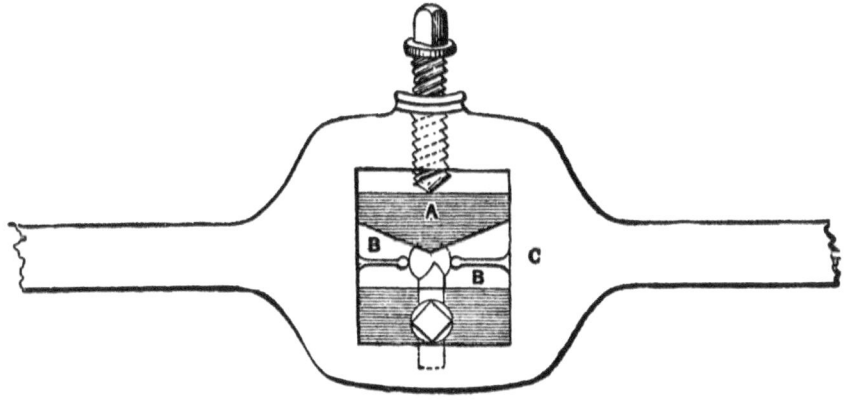

FIG. 146—PLAN VIEW OF "W. D.'S" STOCK.

If many bolts are wanted it would be well to make a pair of blank or soft dies or otherwise a piece of thin sheet-brass between the dies, and the sleeve will be of service. I may state that the thread cut by this tool when well made, is equal to any wood screw, whether made of iron, steel, or manufactured brass. —*By* W. D.

WEAR OF SCREW-THREADING TOOLS.

It is well-known that a tap can be sharpened by grinding the tops of the teeth only, and since the reason of this explains why the work goes together tighter as the tools wear, permit me to explain it. In Fig. 147, which presents a section of the tap, $A B$ are the top corners and $E F$ the bottom corners of the thread.

Now, as the thread is formed by cutting a groove, and the teeth cut the groove, it is evident that $A B$ cut continuously, but as $E F$ do not meet the work until the thread is cut to its full depth, they do no cutting, providing the tapping hole is the right size, as it should be. It follows then that the corners $A B$ wear the most, and like all sharp corners they wear rounding.

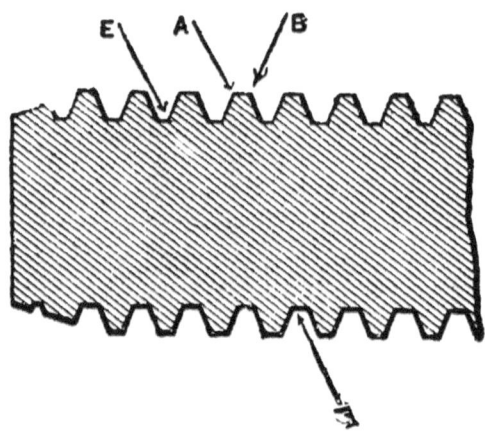

FIG. 147—SECTION OF TAP SHOWING WEAR OF SCREW-THREADING TOOLS.

Now take the case of the die in Fig. 148, and it is evident that the corners C D do the cutting, and, therefore, wear rounding more than the corners $G H$, which only meet the work when it has a full thread cut on it.

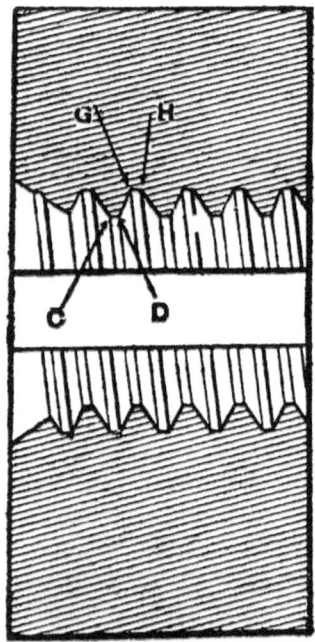

FIG. 148—SECTION THROUGH A DIE.

Now take a bolt and nut with a thread cut on it by worn dies, and we shall have the condition of things shown in Fig. 149; rounded corners at *A B* on the nut and at *C D* on the bolt, and sharp corners at *E F* on the nut and at *G H* on the bolt. Hence corners *G H* jam against corners *A B*, and corners *E F* jam against corners *C D*.

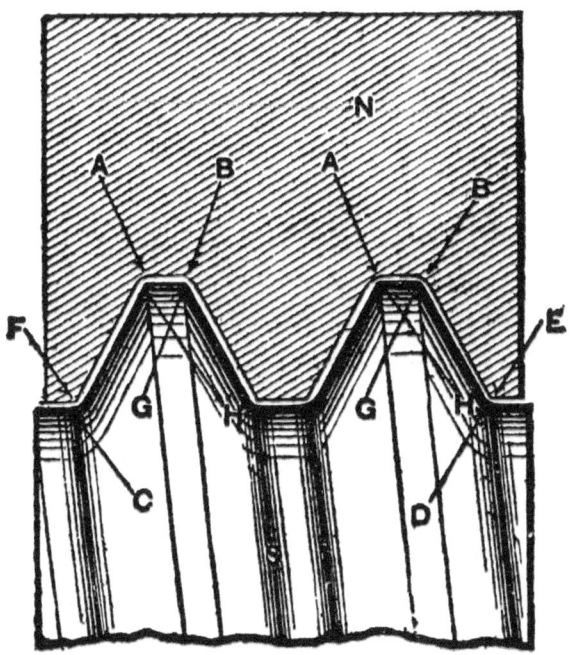

FIG. 149—ENLARGED SECTION THROUGH A BOLT AND NUT CUT BY WORN DIES.

So much for the wear, and now for the sharpening of the tap. It is evident that as the tops of the teeth do the main part of the cutting, they get dull quickest, and therefore by grinding them the teeth are greatly sharpened. — *By* J. R.

TOOL FOR WAGON CLIPS.

The following tool for use in bending buggy, saddle and wagon clips, is easily made and will save much time where a number of clips are used. Figs.

151 and 152 show the tool taken apart. In making it have the width from axle to axle the required width of the clip, and have the part *A* so that it will fit up tight in the slot *H*. The part *B*, shown in Fig. 151, is of the same width as *A*. Fig. 150 shows the tool put together with the clip fastened in it ready for bending. In bending clips in a tool of this kind, they can be formed half round, if desired. Round and square iron may be used as required. In using the tool proceed as follows: Place *A*, of Fig. 151, in the slot *H*, then place the clip in the slot at *O*.

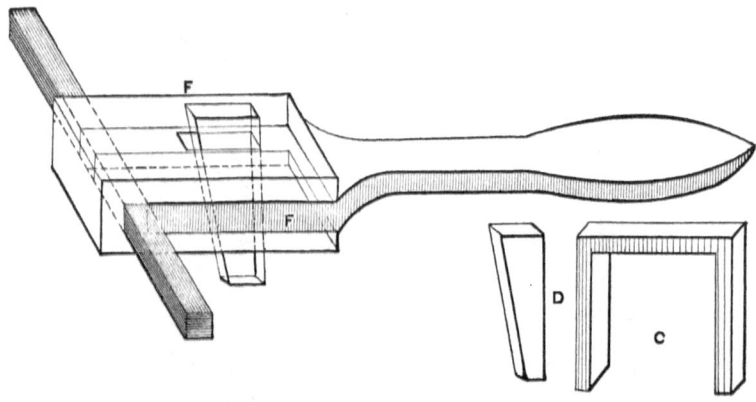

FIG. 150—DEVICE FOR BENDING CLIPS, WITH DETAIL OF WEDGE AND A VIEW OF A FINISHED CLIP.

Next drive the wedge-shaped key *D* in the slots of *C* and *B*.

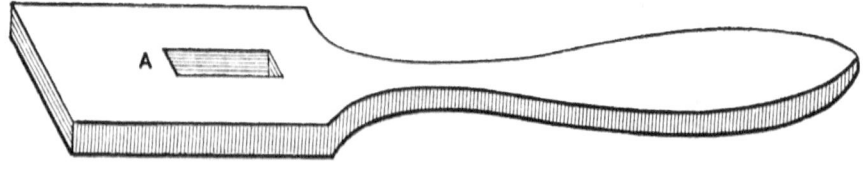

FIG. 151—THE PRINCIPAL PART OF THE TOOL.

When the clip is fastened in the tool, as shown in Fig. 150, with a hand hammer bend the ends against the parts *F F* of the tool. In taking the clip out of the tool drive the pin out of the slots and drive the clevis from the mandrils.

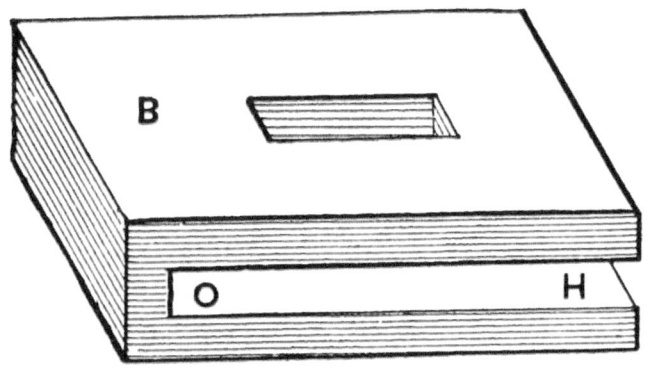

FIG. 152—DETAIL OF THE CLAMPING DEVICE.

At this stage the clip will be finished, or in other words it will be shown in the shape as shown at *C* and ready for swedging the parts for the nuts. By a little practice any smith will be able to bend nearly a square corner on the outside of the clip. —*By* H. R. H.

A HANDY TOOL.

Many times, in taking old carriages apart, considerable trouble arises from the turning of bolts, as the small square under the head will not hold the bolt if the nut turns hard.

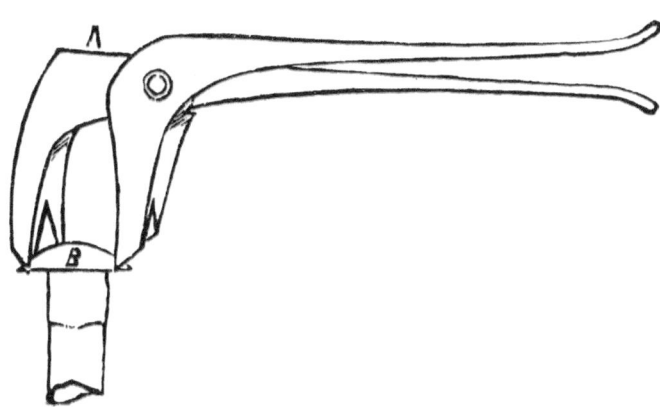

FIG. 153—A HANDY TOOL.

In such cases it is usually necessary to split the nut, to get the bolt out. The tool in question will save many bolts and much vexation of spirit. Set the jaws down over the head B of the bolt, Fig. 153, strike on the top of the jaw at A with a hammer to settle into the wood; then pinch the handles and you have it held fast, and the nut can then be readily turned off. The tool is about 10 inches long, and 5-8 to ¾ of an inch in thickness, measuring through the points joined by the rivet. It will work on bolts from 3-16 to 3-8, and even larger. A blacksmith can make a pair in one hour, and save many hours of valuable time. For the want of a better name we call it a "Polly."—By O. F. F.

FALSE VISE-JAWS FOR HOLDING RODS, ETC.

Here is a simple device for holding bolts in a vise. Though an old invention it is not as well-known as it ought to be.

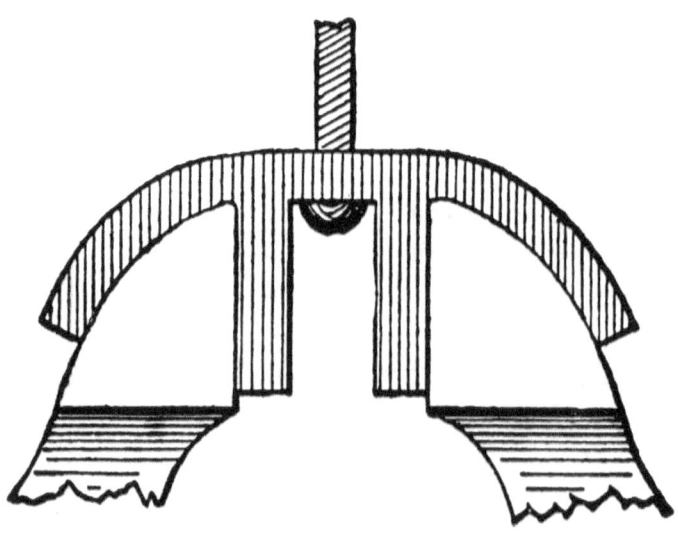

FIG. 154 —FALSE VISE-JAWS. END VIEW.

Fig. 154 of the illustration shows the end of a pair of cast-brass jaws fitting on the vise.

Fig. 155 shows one jaw. The same pattern will do for both. Clamp the jaws and drill the holes, filing them afterward to the shape desired.

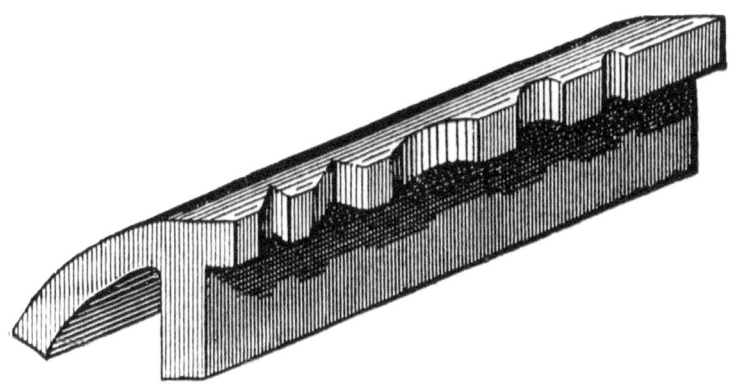

FIG. 155—SHOWING ONE OF THE JAWS.

Square holes are most useful, as they hold both round and square rods. Do not make the holes larger than need be, and then they may be redressed when worn, or altered to suit new work. —*By* Will Tod.

MAKING SPRING CLIPS WITH ROUND SHANKS AND HALF-ROUND TOP.

The accompanying illustrations represent my method of making short spring clips. The bottom tool shown in Fig. 156, should have a handle about 12 inches long. To make this tool take cast steel, say 3x1 inches, and first forge out the handle, then drill the two holes XX.

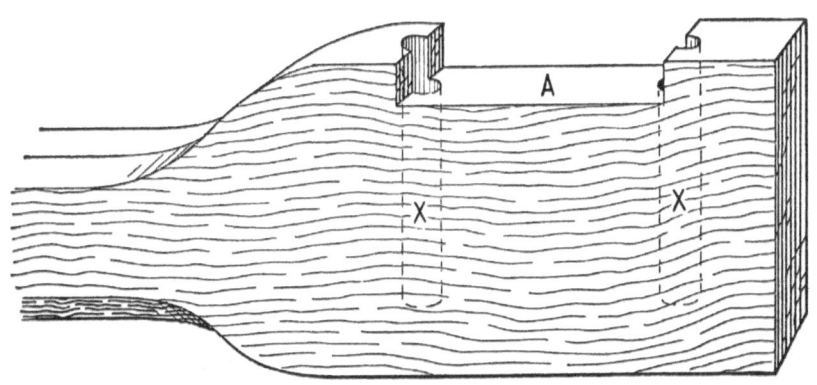

FIG. 156—THE BOTTOM TOOL FOR MAKING SPRING CLIPS.

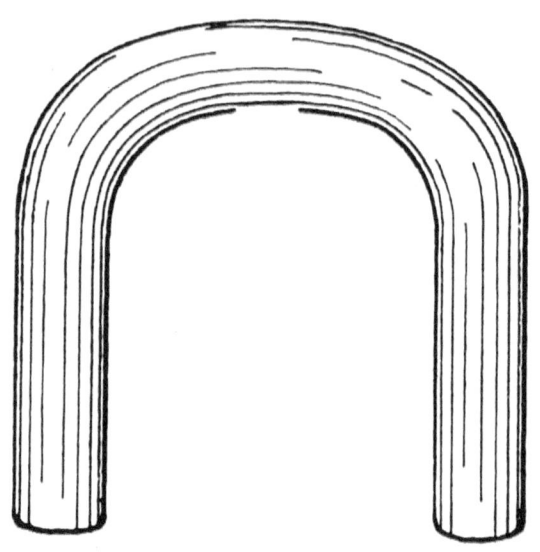

FIG. 157—SHOWING HOW THE CLIP IS BENT.

These holes should be for 5-16 iron, and drilled with a 11-32 drill as deep as the shaft required is long, and should never be drilled through the tool, but made as shown in Fig. 157 by the dotted lines XX. Fig. 159 represents the top swage.

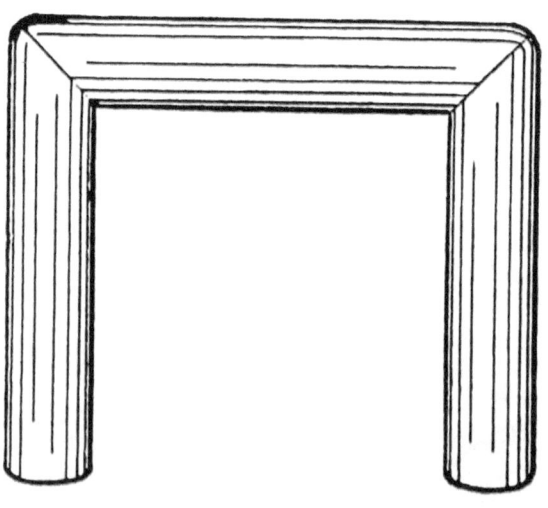

FIG. 158—THE FINISHED CLIP.

This is made of cast steel, the same as any common top swage. First cut out the impression *C* the full length of the tool, then cut down the recess *B B*, then punch out the eye *D* for the handle.

This top swage is made to fit the bottom swage or handle tool, Fig. 156. To make the clips proceed as follows: Cut 5-16 inch round iron the proper length, bend it as shown in Fig. 157, place it in the tool, Fig. 156, set the swage, Fig. 159, on top of the iron, and with five or six blows from the sledge you get the clip with both corners bent, as shown in Fig. 158.

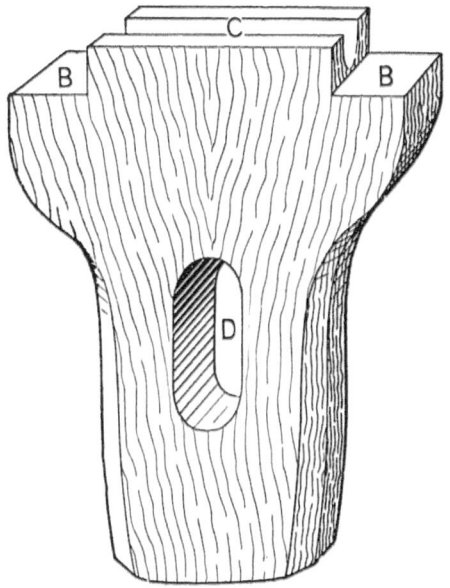

FIG. 159—THE TOP TOOL USED IN MAKING SPRING CLIPS.

Any smith after two or three trials will know just how long to cut the iron to get the best results. The tool must be kept cool, for when it gets hot it will stick the clip. —*By* H. R. H.

HANDY TOOL FOR MARKING JOINTS.

Not wishing to secure everything and give nothing in return, I send you a sketch, Fig. 160, of a handy tool to mark off joints, where one cylindrical body joins another. *A* is a stem on a stand *E*. A loose sleeve, *B*, slides on *A* carrying an arm *C*, hold a pencil at *D*. A piece of truly surfaced wood or iron, *W*, has marked on it the line *J*. Two Vs, *G G*, receive the work *P. G*.

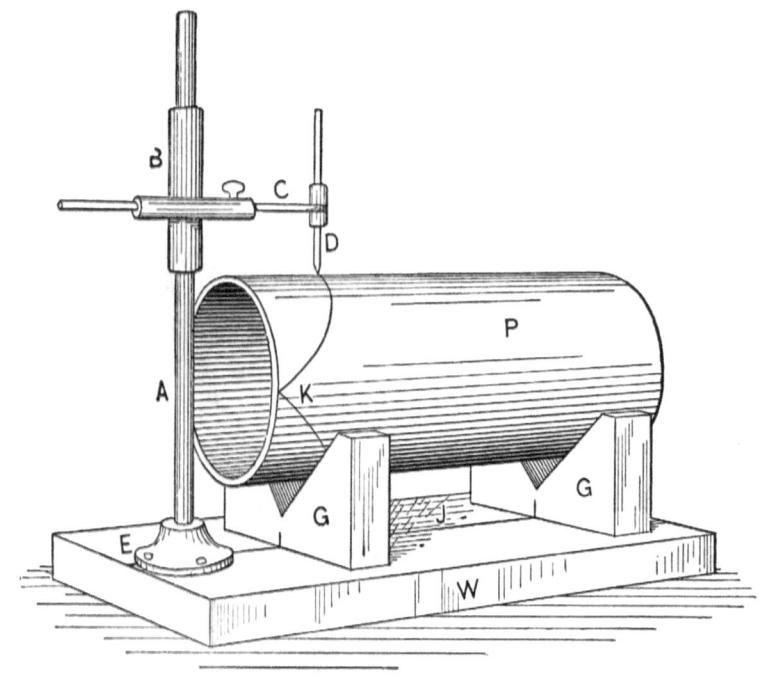

FIG. 160—HANDY TOOL FOR MARKING JOINTS.

Now, if the centers of *G* and of the stand *E* all coincide with the line *J*, then *E* will stand central to *P*, and *D* may be moved by the hand around, *P* being allowed to lift and fall on *A* so as to conform to the cylindrical surface of *P*, and a line will be marked showing where to cut away the wood on that side, and all you have to do is to turn the work over and mark a similar line diametrically opposite, the second line being shown at K. —*By* S. M.

TOOLS FOR HOLDING BOLTS IN A VISE.

I send sketches of what I find a very handy tool for holding bolts or pins. It consists of a spring clamp that goes between the vise jaws, as in Fig. 161 of the accompany engravings, and has a groove in its jaws to hold the round stems of bolts to the vise jaws, without damaging the heads.

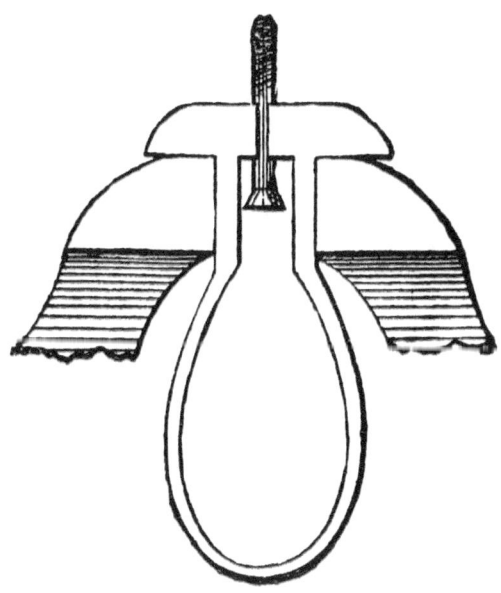

FIG. 161—SHOWING HOW THE BOLT IS HELD BY DEVICE OF "M. S. H."

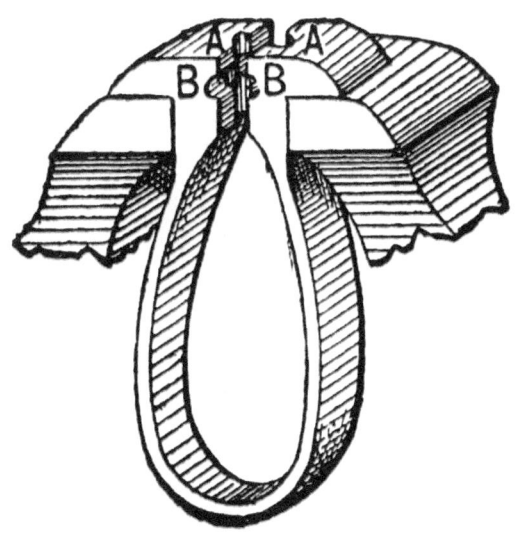

FIG. 162—SHOWING THE DEVICE ADAPTED FOR HOLDING ROUND PINS.

It may be made also, as in Fig. 162, with a hole on its end as well to hold round pins. It doesn't fall off the vise as other clamps do. —*By* W. S. H.

A TOOL FOR MAKING SINGLETREE CLIPS.

I have recently devised a simple tool for making singletree clips from a single heat and will endeavor to explain how it is made.

Proceed the same as in making any kind of a bottom swage, that is, take a piece of square iron the size of the hole in the anvil, and upset one end sufficiently for welding, then take another piece of iron 1/8-inch thick by 2 ½ inches wide and 8 ½ inches long and draw the ends down to the shape shown in Fig. 163 of the accompanying illustrations.

FIG. 163—SHOWING HOW THE ENDS ARE SHAPED.

Upset in the center, or, if your anvil is small, a little to one side of the center, or enough so that when your swage is in place, the end will not project over the end of the anvil. Weld on the stem and fit to the anvil. Next, make two ½-inch grooves in each end of the swage commencing near the end and pointing to the center, as shown in Fig. 164. Have the measurements as near as possible to the following: From *A* to *B*, 8 inches; from C to D, 1 ½ inches.

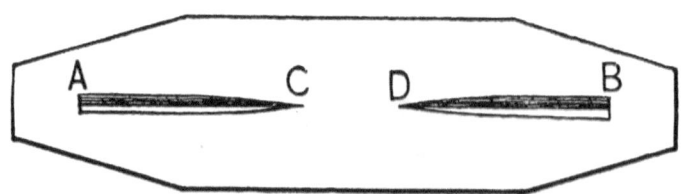

FIG. 164—SHOWING HOW THE GROOVES ARE MADE.

Next, take a piece of iron 1 x 1 ¼ inches and weld on the ends of the swage, extending over the swage about ¼ of an inch. Open up the ends of the grooves and the tool is ready for use, as in Fig. 165.

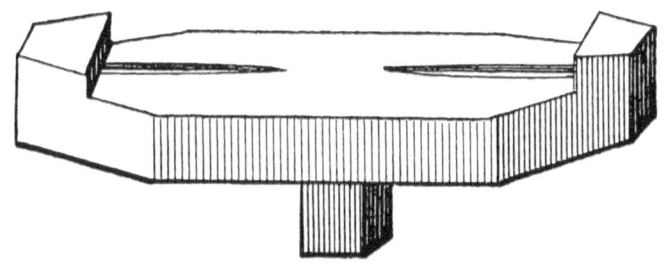

FIG. 165—THE FINISHED TOOL.

To make the clips take ½-inch round iron, cut off 8 ½ inches long, heat in the center and bend sufficiently to allow it to go in the tool, then flatten the center, take the clip out and bend it over the horn into shape. A smith can easily make a clip at each heat. Of course, for making clips of different sizes the tool must be made accordingly. —*By* J. W. C.

Tool for Making Dash Heels.

I send sketches of a tool for making dash-heels for buggies or phaetons, and will attempt to describe the manner of making and using the tool. Fig. 166 represents the article. It is made of cast steel as far as the dotted line shown to the right. At this point is welded on a piece of 7-8-inch round iron for the handle. The tool proper is four inches long, and the handle twelve inches long.

FIG. 166—TOOL FOR MAKING DASH HEELS. GENERAL APPEARANCE OF THE TOOL.

The thickness or depth of the tool is two and one-half inches. It is made wider at the bottom than at the top. Through the center of the tool, as shown in Fig. 166, a hole is made, passing from face to face.

This hole is 7-8 by 5-8 inch, which adapts the tool to making a heel of these dimensions. This weight is ample for a piano box, or phaeton body. The oval cavity in the upper face of the tool is five-eighths of an inch deep.

The manner of using the tool is as follows: Use Norway iron, 7-8 x ½ inch in size, and draw it out wedge shape, as indicated in Fig. 167, heat the end marked *A A*, and place the iron in the tool; let the helper strike it four or five blows with the sledge; next take a splitting chisel and while the iron is in the tool, split it, as shown in Fig. 167 at *O*; replace the iron in the fire, and obtaining a good welding heat, put it in the tool again and hammer down the split ends in the oval part of the tool.

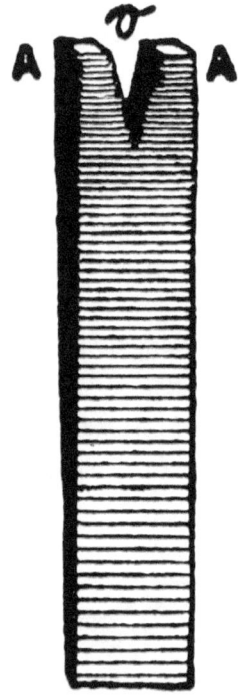

FIG. 167—TAPERING AND SPLITTING THE PIECE OF IRON TO FORM THE ARTICLE.

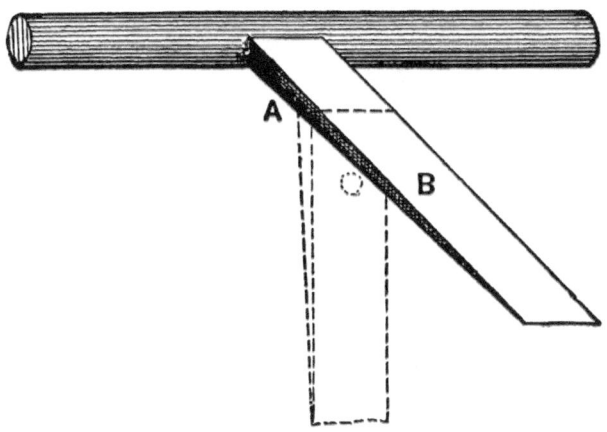

FIG. 168—APPEARANCE OF THE ARTICLE UPON REMOVING FROM THE TOOL.

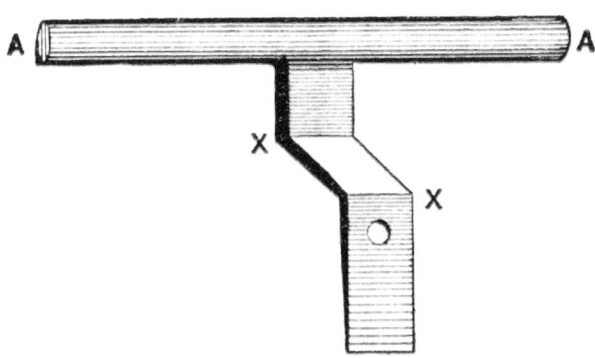

FIG. 169—FORM OF THE ARTICLE MADE IN THE OLD WAY.

With the same heat edge up the iron; this is done by letting the piece come up two or four inches in the tool, and holding it with the tool while edging it up. With the same heat also knock the iron down into the tool and swage with a top oval swage to match the oval of the tool. Take the iron out. It will be found that a heel has been formed as shown in Fig. 168. If the heel is required for a phaeton body, all that remains to be done is to punch the holes; if, on the other hand, however, the heel is wanted for a piano-box or any other body with a panel in the front, it is necessary to bind the corner *A*, Fig. 168, as shown by the dotted lines.

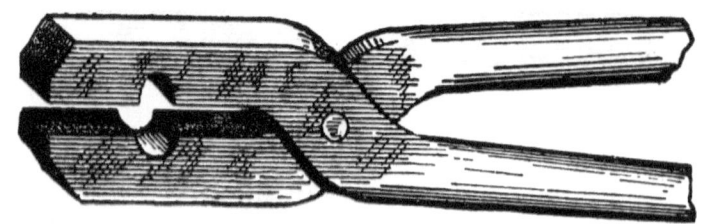

FIG. 170—TONGS FOR MAKING THE BEND SHOWN AT A IN FIG. 168.

The old way of forging a dash heel was to split it, forge and swage the oval ends $A\ A$, Fig. 169, and then bend the corners as shown by $X\ X$ in the same figure. By forging a dash heel with the tool above described, bending one corner is saved, and the piece, when finished, does not set up like a stilt.

The oval iron comes down on to the body. In fastening the dash to the body always let the oval iron project one-eighth of an inch beyond the edge, so as to have the leather of the dash lie straight against the edge of the body. For making the bend A in Fig. 168, I use a pair of tongs shown in Fig. 170. The lower jaw is one inch in depth and the upper jaw 1 1/8 inches. The length of the jaws is four inches, and their width 1 1/8 inches. The manner of using these tongs is so evident upon inspection of the sketch that further explanation is unnecessary. —*By* H. R. H.

MENDING AUGERS AND OTHER TOOLS.

It often happens that a good auger with the screw broken off is thrown away as useless. Now I will try to tell how I have often made a quarter repairing such augers.

Take a file of suitable size and cut a groove the width of the old screw about 3-16 inch deep, a little wider at the bottom than the top (dovetail form). Then form a piece of steel the shape of the screw with a base to it neatly and tightly in the groove. Then coat the edges with a mixture as follows:

Equal parts of sulphur and any white lead with about a sixth of borax. Mix the three thoroughly, and when about to apply the preparation wet it with strong sulphuric acid, press the blank screw tightly in the groove, lay it away five days, and then you will find it as solid as if welded; then smooth up and file the threads on the screw. The job will not take a half hour's work, or cost three cents for material, and the same process may be used for mending almost any broken tool, without drawing the temper. —*By* D. F. Kirk.

An Attachment to a Monkey Wrench.

I enclose sketches of a tool that I have found very useful in my shop. It is an attachment for a monkey wrench. It is made of steel and of the same size as the head of the wrench.

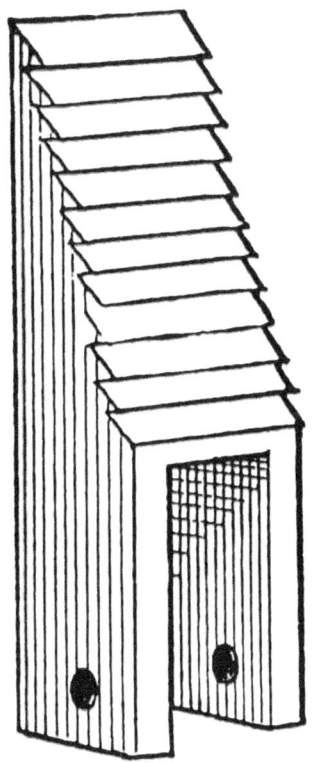

FIG. 171—SHOWING HOW THE TEETH ARE MADE.

The teeth are filed in so that they slant downward toward the wrench, as shown in Fig. 171. A small tire-bolt holds the attachment in place. Fig. 172, represents the attachment in position on a wrench and gripping piece of pipe.

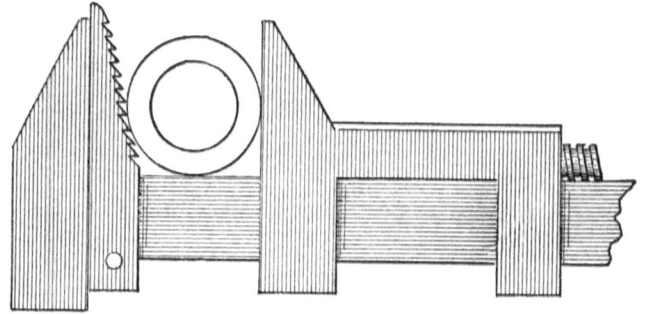

FIG. 172—SHOWING THE ATTACHMENT
IN OPERATION ON THE WRENCH.

This device will hold round rods or pipe as well as a pair of gas-pipe tongs would. —*By* G. W. P.

A HANDY TOOL FOR FINISHING SEAT RAILS, ETC.

Fig. 173 represents a tool that I have found very handy in finishing eyes in seat rails, braces and other work that requires to be fitted exactly. I made this tool like a number of others that I use, for a particular job. It answered the purpose so well that I made others, of different sizes. It can be made by any machinist.

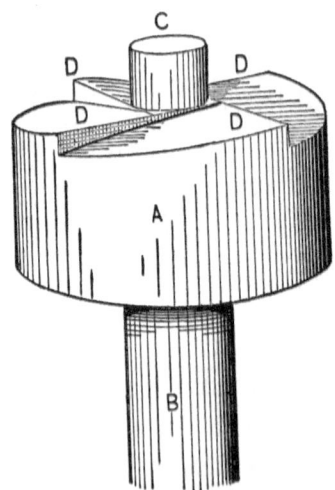

FIG. 173—A TOOL FOR FINISHING EYES IN SEAT RAILS, ETC.

The shank, *B*, is 2-1-2 inches long and 1-2 inch in diameter to fit the drilling machine. The head, *A*, and the follower, *C*, can be made to agree with the work they are to be used on. The cutting lips, *D*, are filed to shape, and tempered to straw color. With this tool you can smooth up. —By "Blacksmith."

A TOOL FOR PULLING YOKES ON CLIPS.

The illustration, Fig. 174, represents a very handy and useful tool of my own invention, which I use for putting yokes on clips or as a clip puller.

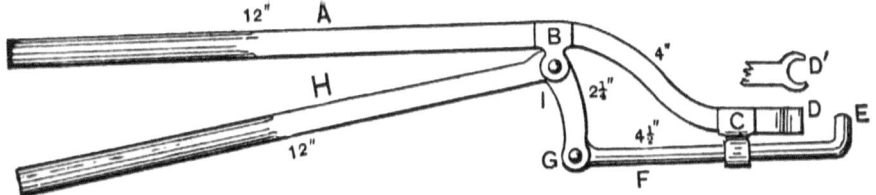

FIG. 174—A TOOL TO PULL YOKES ON CLIPS.

A and *H*, as shown in the illustration, are of 3-8 inch square steel, *A* has two ears welded to it on each side at *B*, and a loop is welded on at *C*; *F* hinges on to *H* at *G*, and *H* hinges on to *A* at *I*; *E* hooks into the yoke; *D* hooks on the end of the clip. By closing the handles, the yoke is pulled to the clip or vise. —*By* A. D. S.

MAKING A CANDLE HOLDER.

I was driving, and full thirty miles from a railroad and three miles from the town I wished to reach, when I lost a jack bolt. Fortunately a little smithy was near, but it was late in the day, and before the son of Vulcan had finished his evening meal and was ready to attend to my wants it was dark. With tallow candle in hand, the smith, with his man and I, went to his shop and the job was done; the smith doing the work, the helper holding the candle. I asked the smith if he found it profitable to pay a helper to hold the candle, and he answered that he knew no better way. I told him that I would pass by his shop again the next day and would show him then how to make a cheap and handy candle-holder. I kept my word and did the job before his eyes, as follows:

I took a piece of band iron, 1 ¼ inches wide and 10 inches long, bent it as shown in the accompanying illustration, so as to join the two angles, each 3 inches long, leaving the back, A, 4 inches. I then bored the holes $B B$, 1-2 inch, and at L (on A) made a 3-8-inch hole.

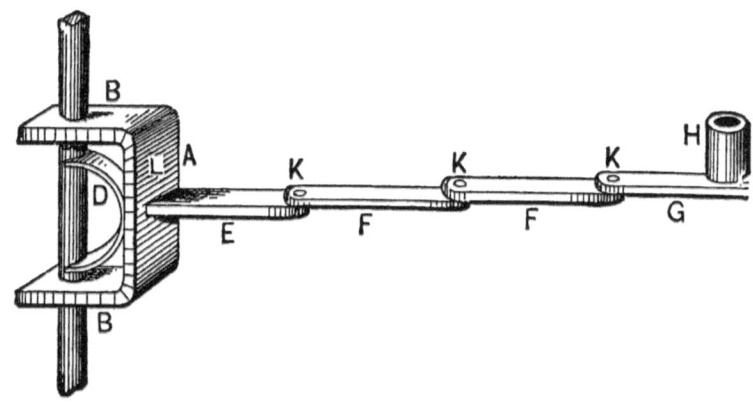

FIG. 175—SHOWING CANDLE HOLDER COMPLETE.

I next took iron ¾ x 3-16-inch and 6 inches long, turned a shoulder on it after upsetting, squared the hole at L and fitted into it the end on E. I next took a piece of 1 x 12 inch hoop iron and made the spring D, concaving the ends, which I riveted on the inner side of A by means of the shoulder on E. I next made the pieces FF and G, and drilled and riveted them as at KKK. I next took a block of wood, H, 1 1-2 inches diameter, 3 inches long, in which I bored a hole to suit the candle, which I secured with a screw on the underside, and my candlestick was complete, as seen in accompanying illustration, Fig. 175.

My friend watched me patiently all the way through, and was inclined to believe that he had me at my wits' end and suggested that he was no better off than before. I asked him to have a little patience.

I soon found a piece of 7-16 round iron, turned an eye on one end, and, making a point on the other, passed it through the bracket. This served as a support. The spring took good hold. I put the stake in the bench, lowered and raised the holder and turned it every way. —*By* Iron Doctor.

MAKING A BOLT TRIMMER.

I have a bolt trimmer which is very easily made, and cannot help giving satisfaction. It is shown complete below. The cutting jaws are of good tool steel. The stationary jaw is welded to the iron handle at A, making the handle and cutter one piece. The movable cutter works in a fork, as seen in Fig. 176.

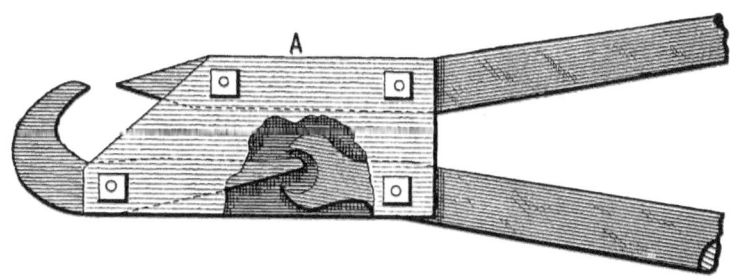

FIG. 176—SHOWING "C. C. O.'S" BOLT TRIMMER COMPLETE.

I have my cutter made with side plates, 3x7 inches, and handles two feet long. For convenience in sharpening I have put it together with bolts.—*By* C. C. O.

A LABOR-SAVING TOOL.

I send drawing of a tool that I have found to be very handy. It is my own idea and a handy tool, especially in plow or buggy work. For plow work it cannot be beaten. The holder is of cast steel and made with two spurs to fit against bolt heads.

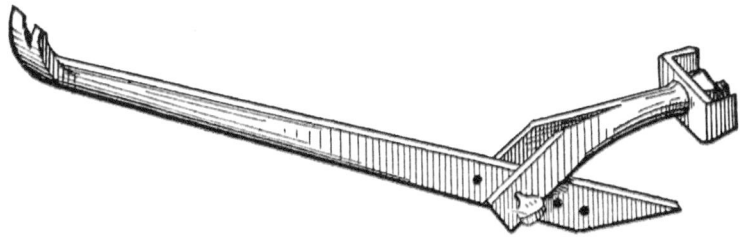

FIG. 177—SHOWING "B. F. C.'S" LABOR-SAVING TOOL.

The steel is welded into a ¾-round iron, 18 inches long. Drill four or five holes for the pin that runs through the holder.

To make the holder take two pieces of 1 ¼ by 5-16 flat iron, weld to make as shown in Fig. 177. Round the end and cut threads for burr brake or pin to hold. When in use the other end is made to pull the bolts to the jaw.

It will be seen from Fig. 177 that this holder is made to turn in any direction and will hold on any shear. —*By* B. F. C.

MAKING A SPIKE-BAR.

I will try to describe a handy spike-bar and tie-fork for mine road men. This tool enables them to do a third more work than they could do with the spike-bar generally used in coal mines. In making the bar I weld 1-inch square steel to 1-inch round iron and use 7-8-round for the fork-prongs.

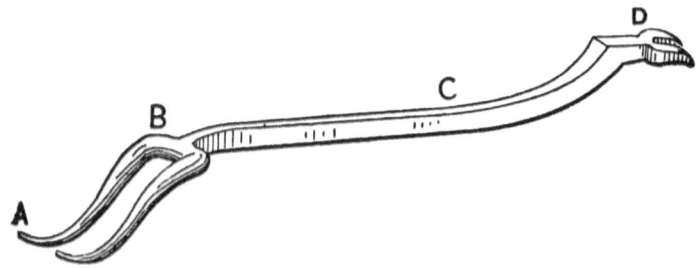

FIG. 178—HANDY SPIKE-BAR AND TIE-FORK FOR MINE ROAD MEN.

A, in the accompanying illustration, Fig. 178, represents the part that goes under the tie, and *B* is where the tool rests on the rail. The man, while driving, sits at *C*. *D* should be made ½ inch longer on the under side than the spike is. —*By* R. Delbridge.

HOW TO MAKE A TONY SQUARE.

This is a very simple thing to make, and very useful for trying six or eight-square timber or iron. It is very handy in making wheelbarrow hubs or anything of that sort.

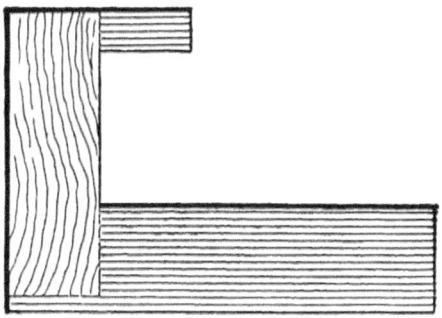

FIG. 179—SHOWING "VILLAGE SMITH'S" TONY SQUARE.

Take an ordinary try square and saw a slit in it opposite the blade; next take a piece of steel plate of the same thickness of the blade and cut it to about one-third of the length of blade in the try square. Insert the short blade in the slit as shown in Fig. 179 and you have a "tony" square that will do its work as nice as you please. —*By* Village Smith.

AN EASY BOLT CLIPPER.

I have a bolt clipper of my own invention and which I think is a very good one. It is simple and easily made The *lever E,* in Fig. 180, and the jaw *I* are of one piece of solid steel. The lever *L* and jaw *M* are also of steel, but in two pieces, as shown by the dotted line at *H*, which is a joint that works by the opening and closing of the handle or lever *L*.

F is a steel plate, there being another on the opposite side. The plates are three-sixteenths of an inch thick. The distance from *A* to *B* is five inches from center to center of bolt holes. From *B* to *C* is two inches from center to center, and from center of bolt hole *C* to joint *H* is three and one-half inches.

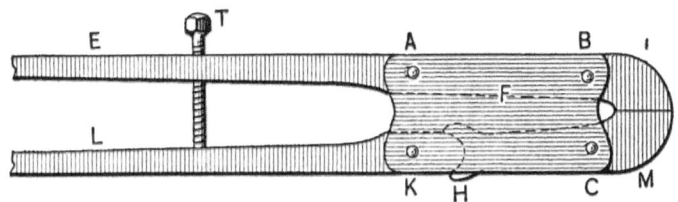

FIG. 180—AN EASY BOLT CLIPPER, AS MADE BY CHRIS. VOGEL.

From joint *H* to center of bolt hole *K* is one and one-half inches. The handles or levers are two feet six inches in length. The set-screw *T* is used to prevent the jaws *I* and *M* from coming together. With this clipper I can cut anything up to a one-half inch bolt. —*By* Chris. Vogel.

A TOOL FOR PULLING ON FELLOES.

The illustration, Fig. 181, shows a tool I have for drawing on felloes when making or repairing wheels. The ring, *A*, which goes over the nut is about 10 inches in diameter; the rod, *B*, is 2 feet long and made from 5-8 iron. It has a thread cut one-half the length. C is a comet nut about one foot long before being bent. *D* is a piece of 1 ½ x 3-8 inch flat iron and bent as shown.

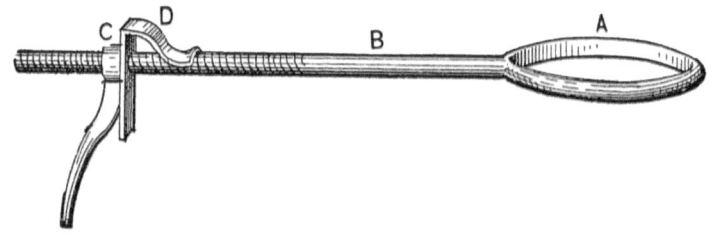

FIG. 181—TOOL FOR PULLING ON FELLOES, MADE BY LOUIS TUTHILL.

It is drilled 11-16 so that it will easily slip over the rod; the end is widened and slightly turned up. There is power enough in this to draw a small wheel all out of shape. —*By* Louis Tuthill.

HOW TO MAKE A HANDY HARDY.

I have a very handy hardy which can be made with little expense. Take an old saw file and break off a short piece from each end. Draw the temper and it is ready for use. If it should stick to the iron when cutting, grind the sides a little. If you are careful to lay it level on the anvil you will have no difficulty in cutting heavy tires—By turning the iron. I have used one of these hardies for a year without breaking it. —*By* R. C.

A HANDY CLINCHER.

The accompanying illustration, Fig. 182, represents a clincher which I believe to be a little better than any I have ever seen. It can be made from an old shoeing hammer.

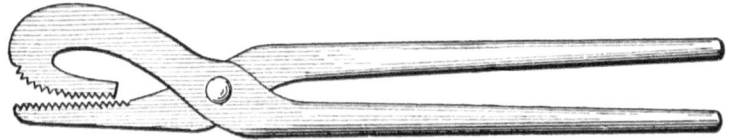

FIG. 182—THE HANDY CLINCHER, AS MADE BY A. F. REINBECK.

The cut shows the construction so clearly that no further explanation is necessary. —*By* A. F. Reinbeck.

A BOLT HOLDER.

I have a bolt-holder that gives good satisfaction, see Fig. 183: it is very useful in preventing bolts from turning. The eye is forged from 7-8 square iron, the eye 5-16 x 1 ½ inches; the part that presses against the bolt-head is steel, the iron being split and spiral inserted; the square part being 5 inches long; from the square to the point, about 4 ¼ inches; the handle welded out of ¾-inch round iron.

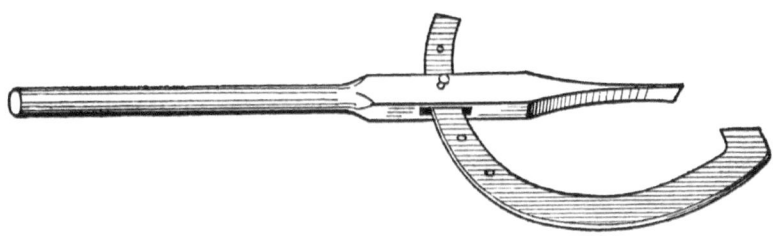

FIG. 183—BOLT HOLDER.

The curved piece passing through eye is made from 1 x ¼-inch stake iron—*By* J. F. Small.

MAKING A CANT-HOOK.

Some smiths may think that the making of a cant-hook is a job too simple to write about, but to make a hook that will catch hold every time is not so easy after all. My way of making such a hook is as follows:

First, make an eye to go around the handle, then make the hook almost any shape, or bend it so that you can then rivet it to the eye and put on the handle.

FIG. 184—CANT-HOOK.

Bend the point so that it will lie flat on the handle when closed, as shown in the accompanying illustration, Fig. 184. Then it will always catch and hold. —*By* E. P. A.

MAKING A CANT-HOOK.

The accompanying illustrations represent my way of making a cant-hook.

The clasp is made of 1 ¾ inch by ½-inch Norway iron. I get the exact measure around the handle, and if it be 9 inches around, I measure 4 ½ inches on the Norway bar (which is of the right length to handle conveniently), then I take a heat, and with a fuller let in about 7-8 inch to 1 inch from each end. I next draw down the center to nearly the right thickness, bend the ends nearly to a square angle, and set down with the hammer, and make the ends or corners square. I then take a chisel and cut in about 1 1/8 inch or 1 inch from one edge, for the jaw, leaving this for the thickest part of the clasp. I then set the remainder down with the set-hammer.

When both ends are down, I draw to the right length, turn and bend to fit the handle.

For the hook, I use 7-8-inch by ½-inch steel, 14 inches long. At the end where the hole is, I upset to make a shoulder, as shown in Fig. 185 of the illustrations, which prevents the hook from cutting away the soft iron of the clasp, and prevents the point of the hook from striking the pick.

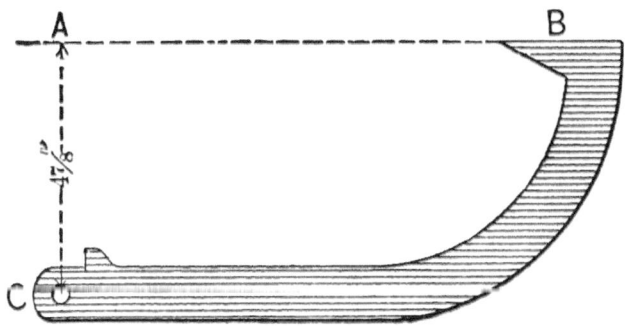

FIG. 185—MAKING A CANT-HOOK. SHOWING METHOD OF MAKING SHOULDER.

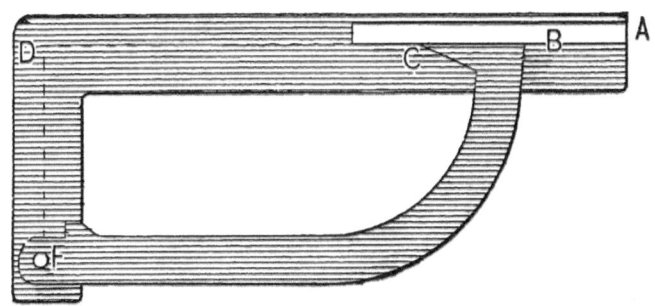

FIG. 186—SHOWING METHOD OF SETTING THE DRAW.

I give 4 7-8 inches draw as shown by the dotted lines *A C*, Fig. 185, and to get this exactly every time, I make the tool shown in Fig. 186. In making it I take a bar, 1 foot 6 inches long, 1 ¾ inches by ½-inch, and bend it 13 inches from *A* to *D*.

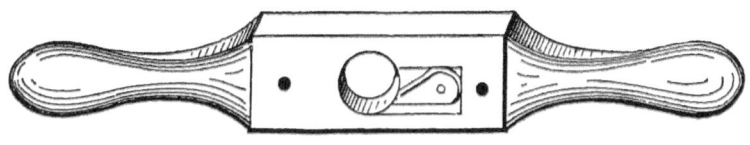

FIG. 187—SCREW BOX.

At B, Fig. 186, I weld on a piece of iron, 6 inches long by ½-inch, with the edge to the *D* bar, and previously bent to the right shape. I then make the piece B, Fig. 186, true on the face (along *C*) with the square *D*. I next measure off from *D* 4 7-8 inches to *F*, and here set a 7-16-inch pin. I make the hook bend and lay as shown in the illustration, Fig. 186, being careful to have the hook true at *B D*. I file to a point from the inside of the point.

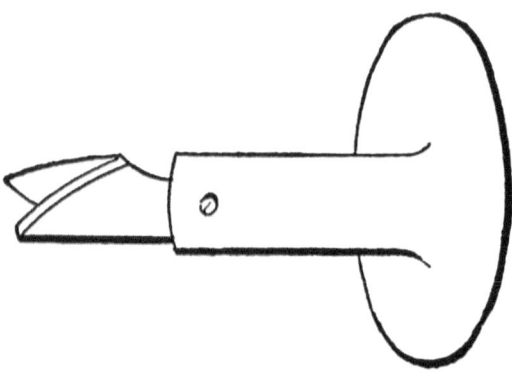

FIG. 188—THE KNIFE.

For the bands, the first is of 1-inch band iron, the two next are of 1 ½ inch, and the toe bands are of 2-inch band iron. This pick is of 7-8-inch square steel, 10 inches long. —*By* W. W. S.

MAKING SCREW BOXES FOR CUTTING OUT WOODEN SCREWS.

To make wooden screws by my plan, first take a square piece of steel and with a three-cornered file make the thread on all four corners of the steel for about two inches. When this is done you will have a tap as seen in Fig. 187.

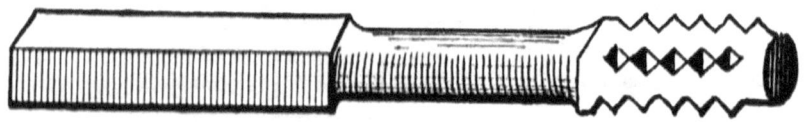

FIG. 189—THE TAP.

To make the screw box as shown in Fig. 188, turn a piece of word (apple wood is the best), with two handles, and bore a hole in the center to the size of the tap with the thread off. Then cut a thread in it with the tap and cut away the wood at one side to admit the knife. This is made as in Fig. 189 with two screws in it, one in the center and the other set.

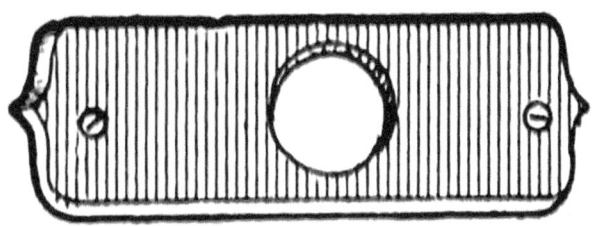

FIG. 190—PIECE OF WOOD USED TO SECURE THE KNIFE IN ITS PLACE AND ADMIT THE TAP.

Put the knife in the box so it will match the thread, and screw in over it a piece of wood one-quarter of an inch thick with a hole in it the size of the tap with the thread on, as represented in Fig. 190. The box is then complete—*By H. A. S.*

MENDING A SQUARE.

In this communication I will tell your readers how to mend a square. Very often a good steel square is rendered useless by having the foot or short end broken off, as in Fig. 191 of the accompanying illustrations. I then work a piece of good iron into the shape shown at A, in Fig. 191, and taking a hack saw, I cut a notch in each piece in which the piece A will fit tightly.

I have a square at hand to ensure accuracy, and then having my coal well charred, I take good clean brass and lay it on. When it begins to get hot I put on borax powdered fine—I can't braze much without borax. When the brass is all melted it is removed from the fire, allowed to cool, and when it is cool the surplus brass and iron are ground off, and the square will then be as good as ever. Copper is about as good as brass to braze with.

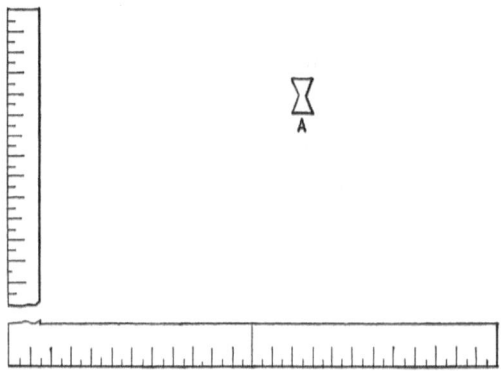

FIG. 191—SHOWING THE SQUARE AND THE PIECE USED IN MENDING.

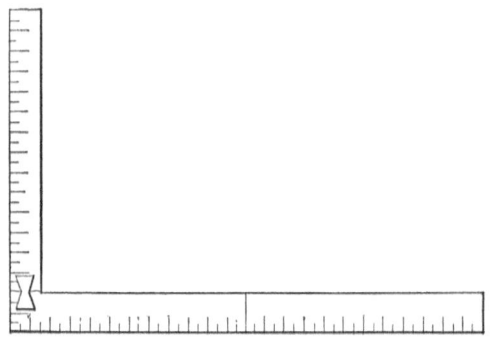

FIG. 192—SHOWING THE SQUARE AS MENDED.

Fig. 192 shows the square when finished. —*By* J. W. J.

STAND FOR CARRIAGE BOLTS.

From an old buggy shaft, three cheese boxes and four strips of wood I made a very handy stand for carriage and tire bolts, the general appearance of which is afforded by the inclosed sketch, Fig. 193. In the center of each box I nailed a square block. I put partitions on two sides, and also two partitions crossways, in order to make six different sized boxes for different sized bolts. I bored a hole through the center and slipped the box down over the shaft.

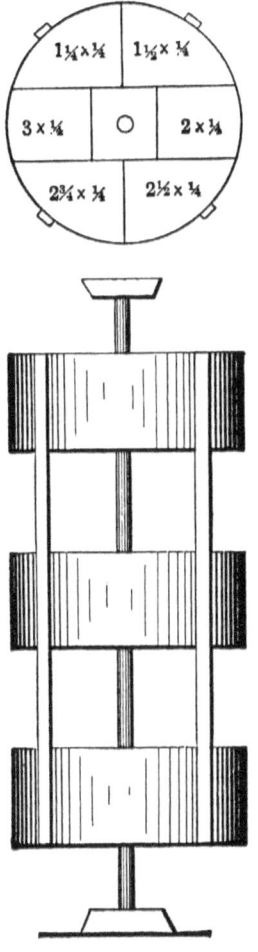

FIG. 193—STAND FOR CARRIAGE BOLTS.

I fastened it both above and below by nails through the shaft. On the outside surface of the boxes I fastened four strips, using ordinary felloe strips for the purpose, placing them equidistant. Their purpose was to keep the boxes steady.

Below and on top I fastened two blocks with holes (bushed) in which the pointed ends of the shaft turn. The device stands in the corner of the shop and is very handy, inasmuch as it easily turns round. Each compartment in the box is marked on the outside in plain figures, thus indicating the size of bolt that it contains. —*By* F. D. F.

AN IMPROVED CRANE AND SWAGE BLOCK.

In the line of cranes I have something differing from the usual style. It is new, I think, and certainly very good. The engraving, Fig. 194, will explain it.

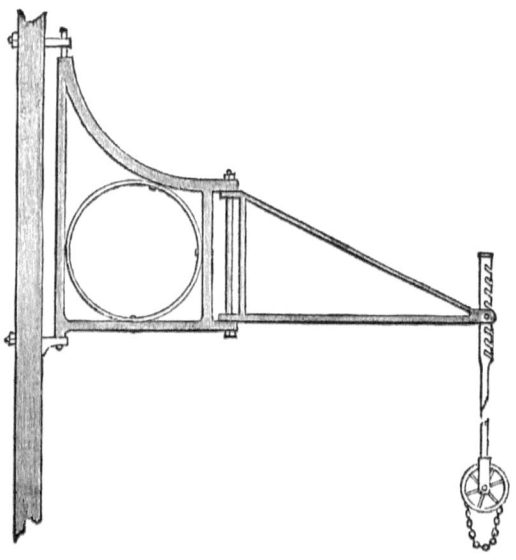

FIG. 194—IMPROVED CRANE.

It can be attached to a post in the most convenient position. Mine is hung in thimbles built in the front of the forge chimney. Next in order shall be the swage block. Of all the much abused tools in a smith's shop, I think the greatest quantity of curses have been bestowed on that patient and unoffending tool. I have known the English language riddled, picked and culled for epithets with the strongest adjectives to hurl at this useful tool. You can hear some of them any time by walking across the shop and stubbing your toe against it as it lays on the floor, and you need not be afraid of hurting it (the swage block I mean). Now, as I consigned mine to the scrap heap many years ago, I will describe a substitute. Get a cast-iron cone mandril, 7 inch diameter at the top and 10 inch at the bottom, with an outside flange at the bottom to form a base, and a strong inside one at the top, having a 4-inch hole in the top, into which cast or wrought iron collets and swages can be fitted for every kind of work, including farmers' and other tools. The cone can be made the height of the anvil and forge, so as to be right for the crane to swing to as easy as the anvil. —*By* Iron Jack.

A CHEAP CRANE FOR BLACKSMITHS.

The accompanying sketch, Fig. 195, of a cheap crane for blacksmiths needs but little explanation, for any practical man will understand it at a glance. *E* is a round pole with a band on each end and a gudgeon and mortise to receive the bar *C*, which is 30 inches x ¾ inch. At *A A* make holes and put in rough pins. Then a part, *B*, is ¾-round iron, with nuts at the top and joint at the bottom.

F is a small sheave, with chain to hold your work, and as you turn your work in the fire or on the anvil it revolves.

I am using one of these cranes, and have had eight hundred pounds on it. In every case it answers well. —*By* Southern Blacksmith.

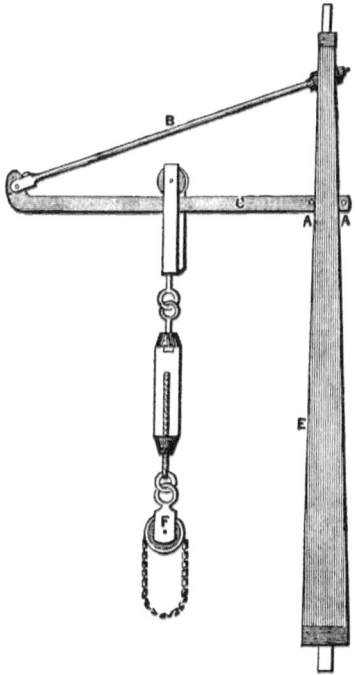

FIG. 195—CHEAP CRANE.

REPAIRING AN AUGER.

I will tell your readers my way of putting a screw in an auger. I take the old auger and file a notch in it where the old screw was broken off. I do this work with the edge of the file, making the notch no wider than the old screw was.

I then take a 3-square taper file and file the notch wider until it appears as at A in the accompanying illustration, Fig. 196. I next take a small piece of steel, forge out the size desired for the screw, file the piece "dovetailing," as shown at B, and then slip it sidewise into the auger. It is put in so it can be driven rather snug. When it is fitted it must be brazed.

FIG. 196—REPAIRING AN AUGER. SHOWING NOTCH AND PIECE TO BE FITTED INTO IT.

Then, commencing inside next to the lip, I file with a 3-square file, and boring the thread half way around, I then commence at the other lip and file a double thread, keeping the two threads side by side and even with each other, by fitting first one a little and then the other about as much, and so on. By this means they can be kept true.

An auger repaired in this way is just as good as new. It does not pay to mend small ones in this way, but it is a good plan for large augers, for the operation is simple and requires but a short time. —*By* Ernest.

A CLAMP FOR HOLDING COUNTERSUNK BOLTHEADS.

I enclose an illustration, Fig. 197, of a clamp that I use in holding countersunk boltheads, while removing the taps from the bolts on spring wagon and buggy felloes. There is no patent on it and it is quickly put on and taken off.

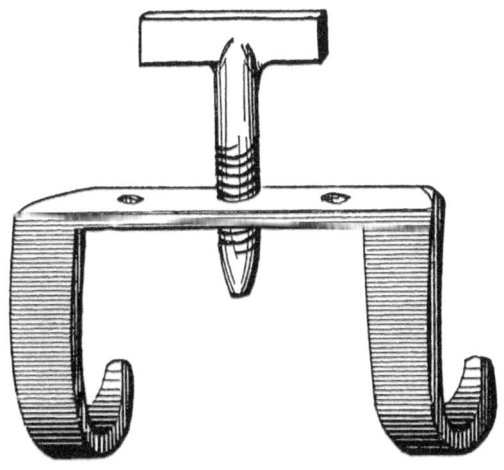

FIG. 197—A CLAMP FOR HOLDING COUNTERSUNK BOLTS.

It is made of 7-16 inch horseshoe bar with three holes, and has a T headed bolt with threads to tighten. The points hook over the felloe and the point of the bolt, which should be tapering at the point, so it will tighten against the bolt without coming in contact with the tire. I have seen different devices for the purpose but like this the best. It should be made from four to five inches long. It will answer the same purpose for a large wheel by making it larger and stronger. —*By* W. E. S.

A HANDY MACHINE FOR A BLACKSMITH.

A useful machine for any blacksmith is made as follows: Take a piece of lumber 1 ½ x 8 and 6 feet long, cut a hole in the middle 2 feet from the end, the dimensions of the hole being 2 x 14, take two cog wheels from some old fan mill, bolt journal boxes for the crank wheel down to the bench on each side of the slot and make an emery wheel mandrel for the small wheel to work on. The mandrel should be of ¾-inch iron 12 inches long. Plug up the hole in the small wheel and bore a hole for the mandrel, having the mandrel square to avoid turning in the wheel, then weld on a collar. If you have no lathe you can true it up with the hammer and file. Next cut a good screw on the end and put your collar on the end, which should be about 2 inches, and put on a small emery wheel ¾ inch thick and 8 or 10 inches in diameter. But first put on a washer of thick leather, also another one against the wheel, screw the tap up tight and if it does not turn true you can trim your leather washer down on one edge and by this means get it perfectly true. On the other end of the bench you can attach a good pair of hand shears. For sharpening drills, cold chisels, and a variety of other work, this machine has no equal. —*By* J. M. Wright.

A CLAMP FOR FRAMEWORK.

The accompanying illustration, Fig. 198, represents a hand clamp for drawing together framework, such as wagon beds, wheelbarrows, etc.

It is made as follows:

The bar A is of narrow tooth steel which will not bend so easily as iron. It should be five feet long by 1 x ½ inch. B is a piece of iron which should be 3 inches high by 1 ½ wide, welded to the end and with a ¾-inch hole having good threads in it. C is a screw to fit the same.

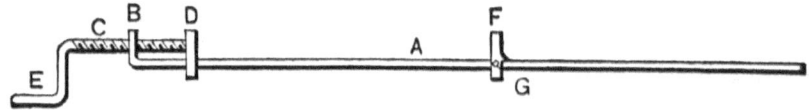

FIG. 198—A CLAMP FOR FRAMEWORK.

It is made one foot long with a crank *E* which is attached to the end. *D* is a slide to fit over *A*, and it should have 3 inches above it a hole ¼ inch deep to allow the screw *C* to get a good bearing. *F* is made the same as *D*, except that it has a shoulder back of it to keep it from leaning too far back, and a set screw *G*, at the side, to hold it stationary. I use this clamp almost every day, and I never saw or heard of one just like it. —*By* V. D. B.

A TOOL FOR HOLDING BOLTS.

I send a sketch, Fig. 199, of a tool for holding loose bolts while screwing nuts off. To make it, take a piece of 5/8-inch round iron of suitable length, draw down oval and tapering about 5 inches, and about 7 inches from the pointed end drive in a piece of steel, wedge-shaped, weld securely and sharpen like a chisel; one inch is long enough for this.

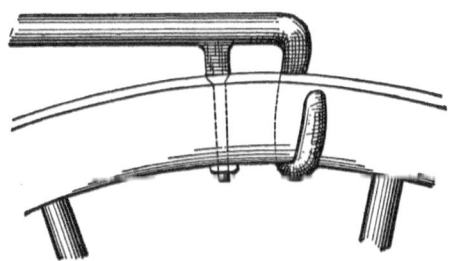

FIG. 199—A TOOL FOR HOLDING LOOSE BOLTS.

Then five inches from the end turn it down at right angles, edgewise, and then curl to the left as shown in the illustration. This is better than all the patented tools for this purpose. —*By* Edwin Clifton.

A HINT ABOUT CALLIPERS.

Let me give some of your young readers a hint how to chamfer off the ends of their callipers from the outside and slightly round them across as in Fig. 200, and not make them rounding as in Fig. 201,

 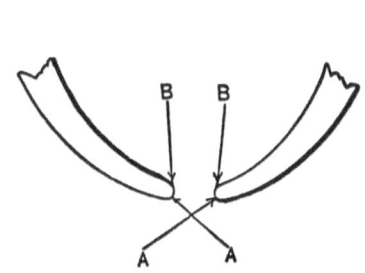

FIG. 200—RIGHT WAY TO SHAPE CALLIPER ENDS.   FIG. 201—WRONG WAY TO SHAPE CALLIPER ENDS

The outer points will always touch at the same point no matter what the diameter of the work. If rounding they will touch, for small work, at A, A, and for large work at B, B. —*By* Shafting.

VISE ATTACHMENT.

I inclose a paper model of a device that I am using for holding beveled edge iron for filing. It is also useful for chamfering flat iron. In use it is to be screwed in a large vise. The spring shown in the cut, Fig. 202, throws the jaws apart when the vise is released. I think many of your readers will find this idea useful, and as it is one that every blacksmith can put into practical operation, I commend it to the attention of my fellow craftsmen. —*By* E. M. B.

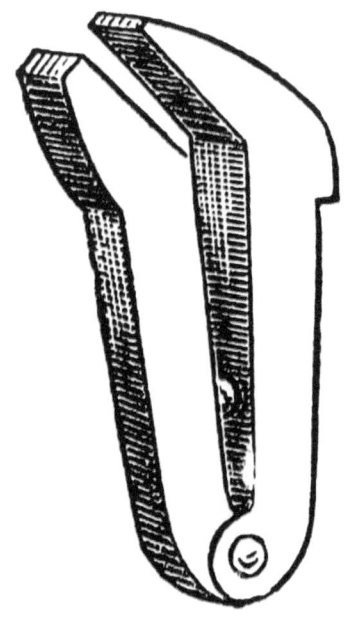

FIG. 202—VISE ATTACHMENT.

Note. —The accompanying engraving has been made from the paper model inclosed in our correspondent's letter, and, we believe, correctly represents his idea. As he did not show how the spring was attached, or in what form it was to be made, we have nothing to govern us in this particular.—Ed.

In beginning to make the lathe, I take a piece of flat iron 12 inches long, 3 inches wide and ¼ inch thick, and cut 3 inches at each end, tapering down to 1 ¾ inches, as shown at a, Fig. 206. I then turn 3 inches of the same ends up at right angles, as at *a*, Fig. 204, and drill two 3/8-inch holes at *b* to bolt the head stock.

The head stock is braced at *c* to prevent the springing of the back end of the frame, as all the end pressure comes on that end. I next drill a ¾-inch hole through the back end and 2 ½ inches from the bottom *a*, as shown in Fig. 206 at *b*, and fit to *b* a piece of round iron 1 ½ inches long, with one end countersunk as in Fig. 207 at *a*. This is to fit the spindle and take up the wear. To prevent this piece from coming out, I double a piece over the end at *a*, Fig. 205.

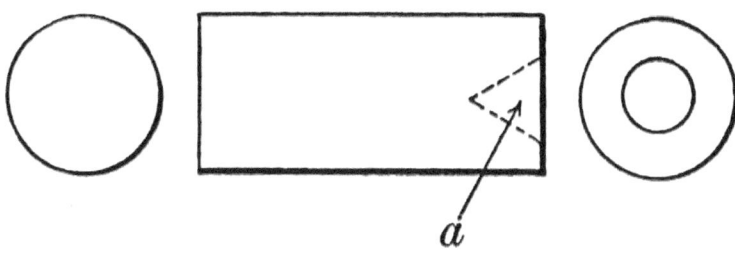

FIG. 207—SHOWING THE PIECE TO BE ATTACHED TO THE SPINDLE.

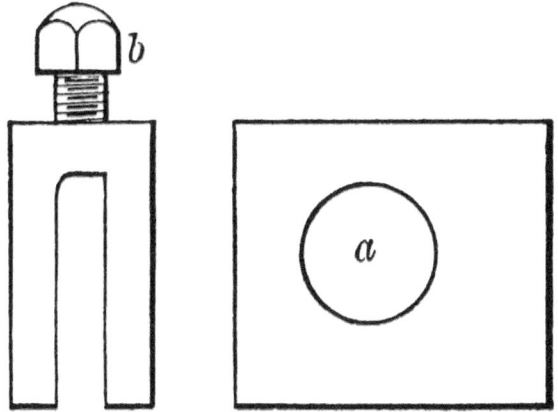

FIG. 208—SHOWING THE PIECE USED TO SECURE IN POSITION THE PART SHOWN IN FIG. 207.

This piece is 1 ¾ by 1 ¼ inches, with a ¾-inch hole, as shown at a, Fig. 208. It has a ¼-inch set screw at b. This piece goes over the end a, Fig. 206, and the piece shown in Fig. 207 goes through the ¾-inch hole, and the set screw bears on the head stock. By turning up the set-screw the piece, Fig. 207, can be clamped at any place desired, thus forming the bearing for that end. The front end has a place cut out at the center, 1 ½ by 1 inch, to receive the boxes.

The edges at a, Fig. 205, are beveled to a V, so the two boxes will slide down and fit tightly. The boxes are 1 ½ x 1 ¾ x 1 ¾ inch. With the ends cut out to fit the V shown in Fig. 205 at a, I next drill in each prong at b, cut a thread and fit a bolt to clamp the boxes. C, in Fig. 209, is the plate that goes over the boxes.

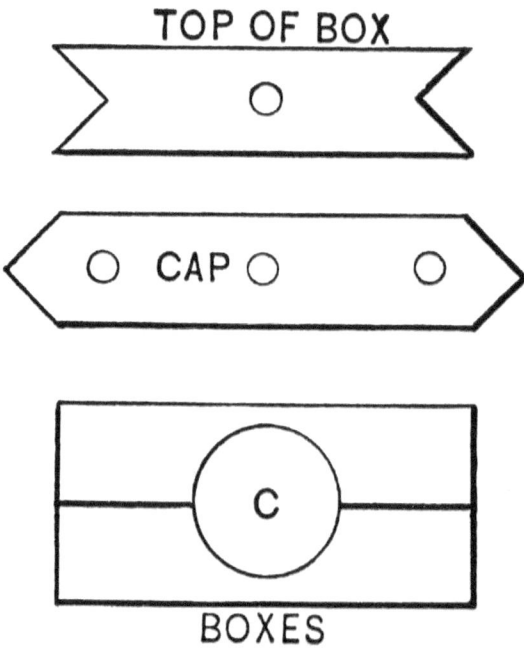

FIG. 209—SHOWING THE BOXES AND PLATE.

The bolts go through the plate into b, Fig. 205. I put the boxes in, placing a thin piece of pasteboard between them, and then clamp them tightly and drill a ¾-inch hole through them at C. Composition is the best material for the boxes.

The spindle must be turned, for it could not be filed true enough to run well. Fig. 210 represents the shape. The end a should fit into Fig. 207 at a.

The bearing at the other end is at b, ¾ inch in diameter. c is turned down a little smaller, and a thread cut on it so as to screw on the face plate. The spur center goes into the spindle with a taper. You can shrink a flange on the spindle at d, and bolt the pulley to that. The face plate needs no description.

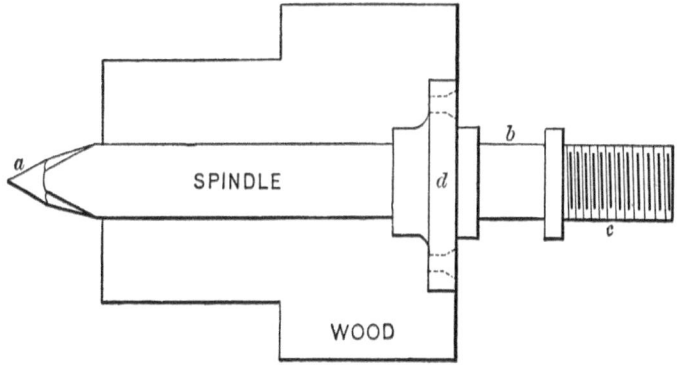

FIG. 210—THE SPINDLE.

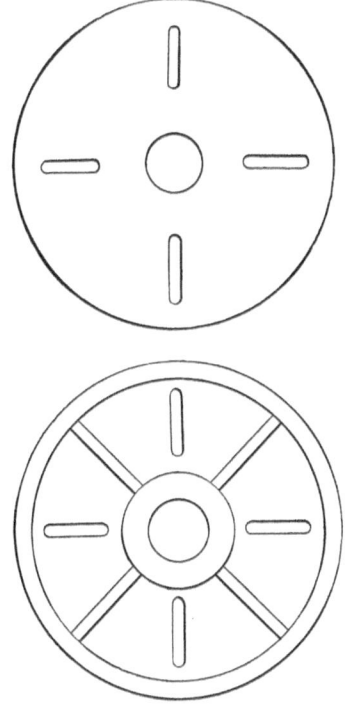

FIG. 211—THE FACE PLATE.

A glance at Fig. 211 will give anyone a clear idea of it. It might be 5 inches in diameter, and it would answer well enough if it were 3 inches only. The tail stock is of the same dimensions of the head stock, that is, 3 inches wide, 6 inches long, and 3 inches high.

Fig. 212 is a side view of the tail stock; Figs. 213 and 214, end views; in Fig. 215 is shown a piece 1 ½ inches by 9 inches by ¼ inch thick, with 1 ½ inches of both ends turned at right angles to *a*. This goes over the ends of Fig. 212 at *A*.

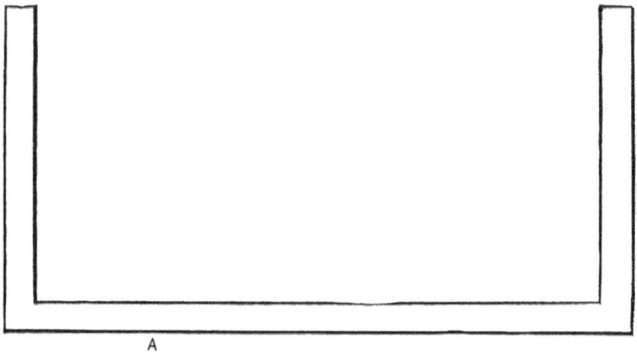

FIG. 212—SIDE VIEW OF THE TAIL STOCK.

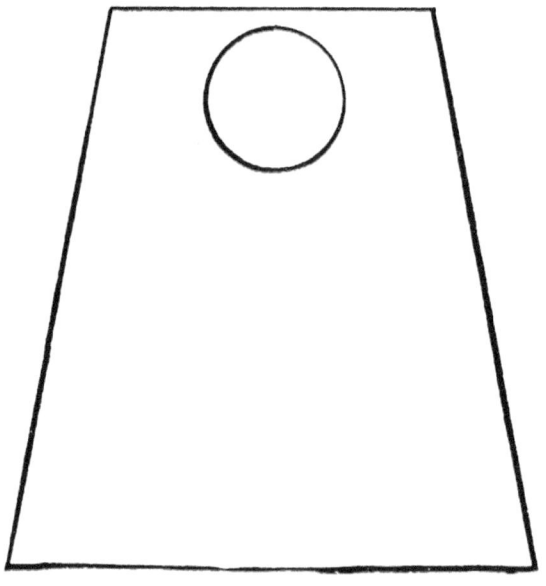

FIG. 213—END VIEW OF THE TAIL STOCK.

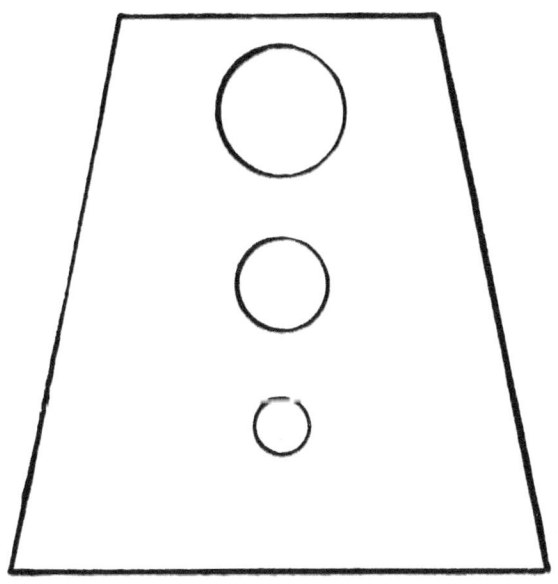

FIG. 214—BACK END OF THE TAIL STOCK.

To clamp the arbor, drill a ¾-inch hole in both ends of Figs. 213 and 214, 3 ½ inches in front of *a*, and going also through the ends of the piece shown in Fig. 216 at *b*. This hole must match the holes in *s*. 213 and 114. The purpose of the arrangement is to hold the arbor and keep the work from coming out of the lathe.

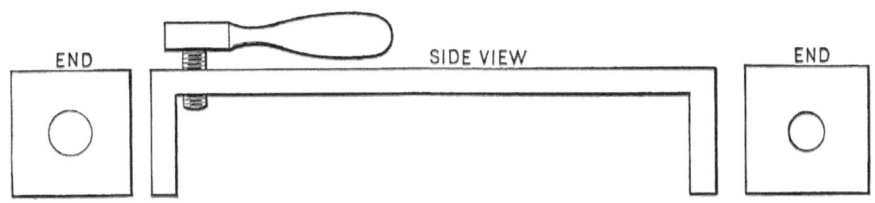

FIG. 215—THE PIECE USED OVER THE ENDS OF THE PART SHOWN IN FIG. 212.

The set-screw shown in Fig. 215 bears on *A*, Fig. 212. When the screw is turned, it will keep the arbor from slipping. Fig. 214 has a ½-inch hole with a thread cut in it.

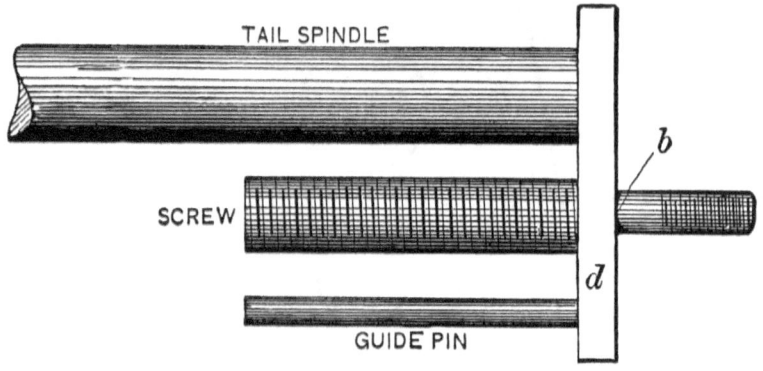

FIG. 216—SHOWING THE TAIL SPINDLE, SCREW AND GUIDE PIN.

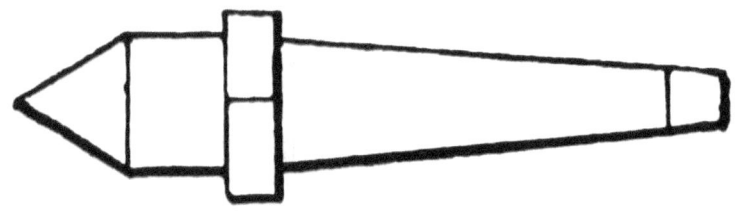

FIG. 217—SHOWING THE CENTER.

There must be another hole ¼ inch in diameter, as shown in the engraving. The arbor is 9 inches long, ¾ inch in diameter, with one-fourth of one end turned down to ½ inch diameter, as shown in Fig. 219. The center goes in the end with a taper as shown in Fig. 217. The center has a place left square to receive a wrench in order to take it out of the arbor. Fig, 218 is a piece 2 ¾ x 1 ¼ inches with three holes in it, one ½ inch, one 3/8 inch, and the other ¼ inch.

These holes should correspond with the three holes in Fig. 214. Fig. 218 is riveted to the arbor, which is worked with a screw. The guide-pin is fastened to the plate and goes through the smallest hole in the piece Fig. 214.

Fig. 220 is a hand wheel which fits on to a very tight nut. To fasten it, there must be work in the plate, so that the screw can be turned in and out. In turning the screw so, you carry the arbor with it. The rest is a flat piece of iron ¼ inch thick, 8 inches long and 3 inches wide, with 2 inches of one end bent at right angles.

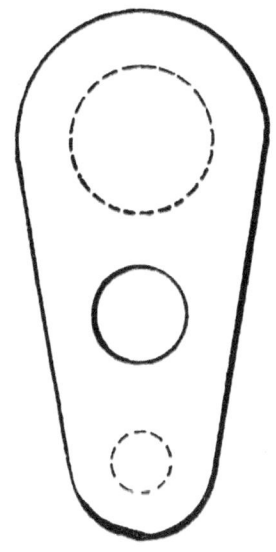

FIG. 218—SHOWING THE PIECE RIVETED TO THE ARBOR.

FIG. 219—THE TAIL SPINDLE.

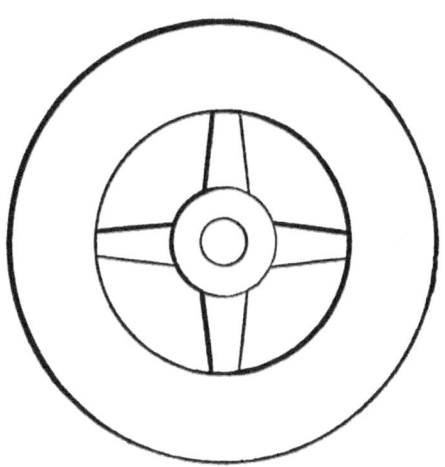

FIG. 220—THE HAND WHEEL.

There must be two holes near the end, so a piece of wood can be bolted on for turning different lengths. To fasten the rest to the bed cut a hole the size of the bolt, 4 inches long, in the bottom of the rest to let it slide to and from the work. —*By* H. A. Seavey.

CHAPTER VIII.

BLACKSMITHS' SHEARS.

I enclose a sketch, Fig. 221, of a pair of shears to be used in the square hole of the anvil. They are very useful and cheap. Any blacksmith can make them. Use good steel and make the blades eight inches long, measuring from the rivet.

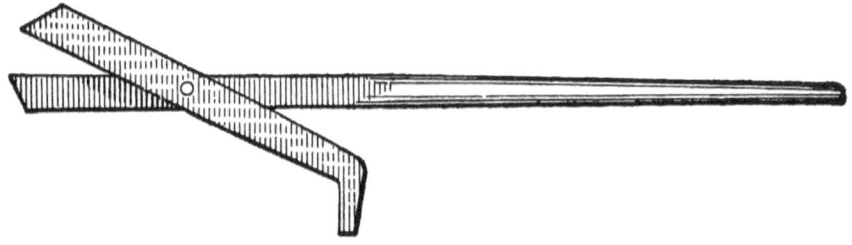

FIG. 221—BLACKSMITHS' SHEARS.

Make the short blade with a crook, as shown in the illustration, to go in the anvil, and have the long blade extend back about two and a half feet to serve as a handle. With these shears I can cut quarter-inch iron with ease and cut steel when it is hot. —*By* A. J. T.

SHEAR FOR CUTTING ROUND AND SQUARE RODS.

I would like to give a description of a shear for cutting round and square iron constructed by me. The inclosed sketch, Fig. 222, is an attempt to represent it. The lower member of the shear is a bar an inch thick, three inches wide and fourteen inches long, and is furnished with a steel face at that part where the cutting is done. The upper member is of the same general description, except that it is seventeen inches long. The lower blade is fastened to the bench at the back part by cleats, as shown in the drawing. A guide for the upper blade, just wide enough in the opening to allow of easy play, is made to serve a like purpose for the front part.

FIG. 222—SHEAR FOR CUTTING ROUND AND SQUARE RODS.

The handle of the shear is hinged to the lower blade, and is connected also with the upper one by the link shown in the sketch. The handle is five feet long and is one by three inches in size down to a taper. Three holes are provided in it for connecting the link attached to the upper blade, thus opening the shear more or less as may be required. With this shear I can cut round or square iron up to seven-eighths in size. —*By* Southern Blacksmith.

CHEAP SHEARS FOR BLACKSMITHS' USE.

I inclose a sketch, *Fig. 223*, of a cheap shears for smiths' use, and submit the following directions for making: The under jaw, *D*, should be 10 inches long, 3 inches wide and 1 inch thick.

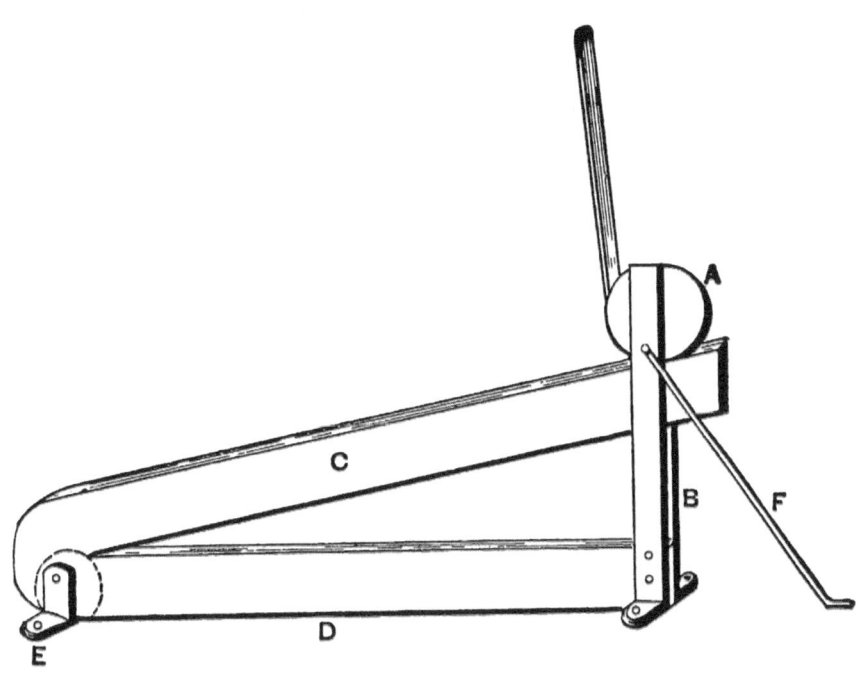

FIG. 223—CHEAP SHEARS FOR BLACKSMITHS.

The upper jaw must be 13 inches long, but otherwise the same as the lower jaw, except where it couples with the latter. Then it must be forged by the dotted lines. The coupling at E is made with a 7/8-inch cast-steel bolt, which takes a brace on each side of the shears, this brace taking one half-inch bolt at the foot through the bench. The braces at the other end take two bolts through the bench. That next to the lower jaw takes two half-inch rivets through the same and a ¾-inch cast-steel bolt at the top through the cam. The upper jaw is brought up by two strips of sole leather connected to the cam A by two bolts. The two braces, F (only one of which is shown in the cut), are ½-inch round and take a 3/8 bolt at the foot. The material for jaws should be 5/8 x 3 inch Swede's iron with the same amount of cast steel or English blister laid on the cutting side, and when finished should have just bevel enough to give a good edge. E and B are made of Swede's iron ½ x 3 inches. The cam, A, is the same thickness as the jaw and finished with 1 inch round for a lever 3 feet long. The jaws should be brought to a low straw color in tempering. The cam must be finished smooth and the bearings kept well oiled.

Then you have a pair of shears at a nominal cost that will last a lifetime and work better than most of the shears in the market. It is a good plan to use a guard with the shears; let it bolt on to the bench, rising 3/8 of an inch above the edge of the lower pair, and then run parallel with the jaw to the other end, where it is secured by another bolt. The brace, B, which rivets to lower jaw, must have an offset of one inch to come flush with the inside of the jaw. —*By* J. M. W.

BLACKSMITHS' SHEARS.

I send a sketch, Fig. 224, of shears made by myself. They are cheap and I have found them very convenient. The engraving from my design requires no explanation. A glance at it will be sufficient for any smith who understands his trade.

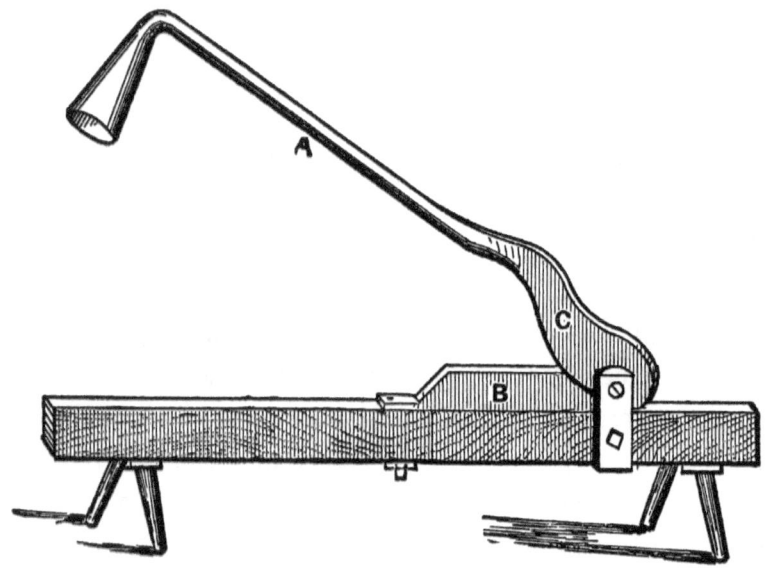

FIG. 224—BLACKSMITH'S SHEARS.

I will, however, give some of the dimensions. *A* is 1 ¾ round, *B* 9 x ¾ and *C* 6 ½ x ¾ inch. The main point in making is to get the edges to come together as in the common shears. —*By* J. J.

SHEARS FOR THE ANVIL.

I send you a sketch, Fig. 225, of a very handy tool, a pair of shears for the anvil. Any blacksmith that understands his trade can make them. They are good for trimming cultivator shovels when they have just been painted and they will take the place of a helper on many jobs where striking is needed.

The cutting jaws are 4 inches long, 3 inches wide and ¾ inch thick, and bevel to the edge and to the back.

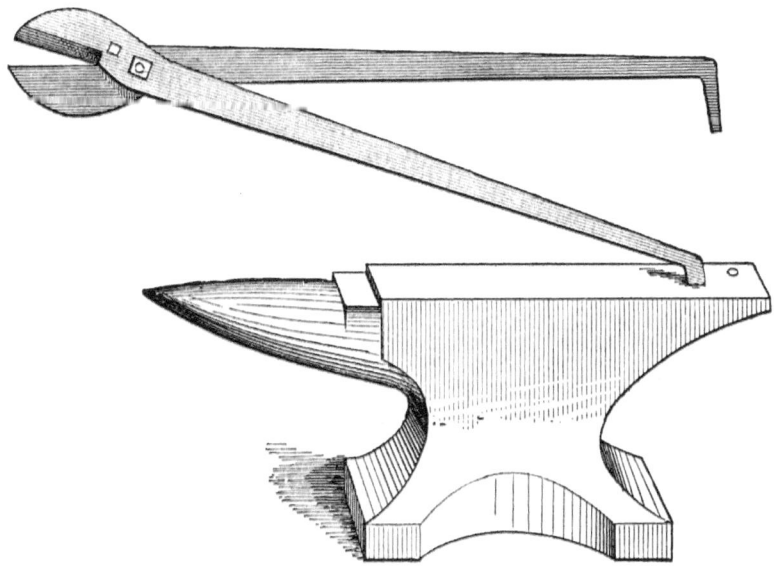

FIG. 225—SHEARS FOR THE ANVIL.

One jaw has a square hole for a square shoulder bolt. The handles are two feet long. I use them on hot iron or steel and they cut sheet iron cold. —*By* G. W. P.

CHAPTER IX.

EMERY WHEELS AND GRINDSTONES.

EMERY WHEELS.

I have polishing wheels in daily use, and put the emery on them with good glue. The way I employ the glue is as follows: I heat it to the proper degree, and then with a brush I cover from six to eight inches of the wheel with it. Then I put the emery on the covered part, and with a roller run over it so as to pass the emery down into the glue. I then apply the glue for another six or eight inches and repeat the same operation. I keep on in this manner until I get around the wheel. I then lay it away for twenty-four hours to dry, after which time it is ready for use.

In making emery wheels, nothing but the best glue is satisfactory for use. Poor glue is worse than nothing. Care must always be taken to keep oil and grease of all kinds from getting on the wheel. I have had some trouble with wheels of this general character, but I have always found the fault to be poor glue or oil that squirted from the shafting on the wheel. I make it a rule always to wash and clean the wheels in warm water when I find them greasy, and then let them dry, and put the emery on anew as above described. By following this plan I have always met with good results. No glue, however good, will hold emery or other parts together when the surfaces to which it is applied are oily or greasy. —*By* H. R. H.

MAKING AN EMERY WHEEL.

It will not pay to put emery on wooden wheels because it flies off in pieces. I know this from experience. It is better to use felt that is made for the purpose. I use felt about 4 inches wide and 1 inch thick. I make a wooden wheel of about 12 inches diameter and 4 inches face and nail the felt on it with shingle nails at intervals of 1 ½ inches. I then drive the nails half way through the felt

BOLT SET

We have been using a tool in this community for a long time, which can be applied to wheels very quickly. Any blacksmith who can make a pair of tongs can produce it.

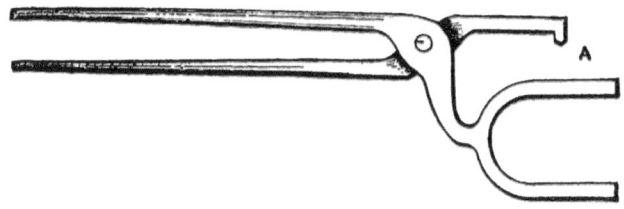

FIG. 203—BOLT SET.

It is made of good steel, A in the engraving being chisel-pointed and hardened, so that it can be set into the head of bolt, when it is necessary, by a slight rap with the hammer. —*By* W. H. S.

A HOME-MADE LATHE.

The accompanying drawings represent a turning lathe that I have been using for some time and find very convenient, not only in turning, but also in drilling small holes. Fig. 204 is a side view of the head stock, and Figs. 205 and 206 show the front and back ends of Fig. 204.

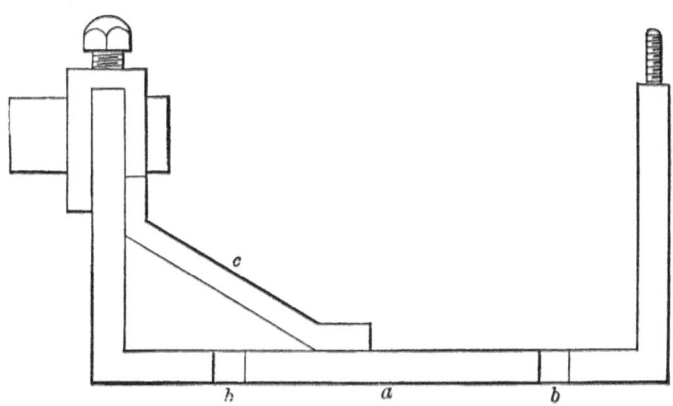

FIG. 204—SIDE VIEW OF THE HEAD STOCK.

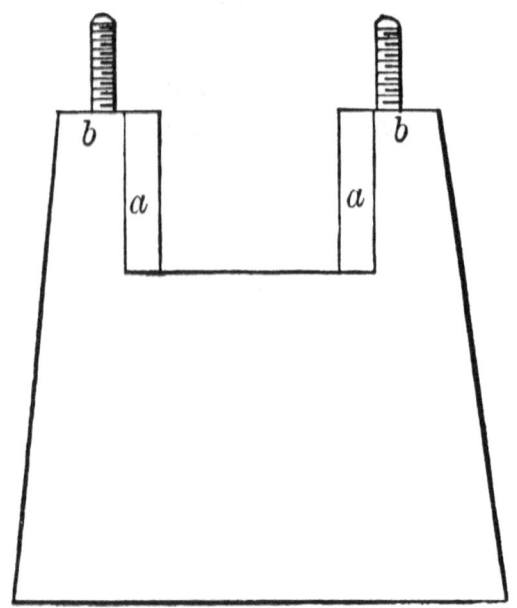

FIG. 205—FRONT END OF THE HEAD STOCK.

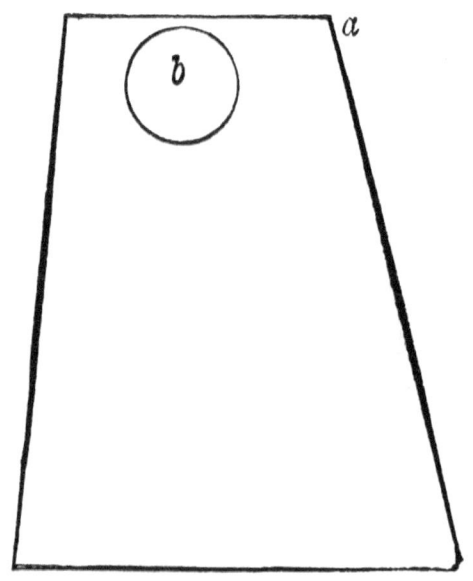

FIG. 206—BACK END OF THE HEADSTOCK.

by means of a punch, spread glue over the felt and roll the wheel in emery. This makes a good wheel for finishing off. —*By* "Shovels."

HOW TO MAKE SMALL POLISHING OR GRINDING WHEELS.

The general method of making small polishing or grinding wheels is to glue together pieces of wood, making a rough wheel, which, when dry, is put upon a spindle or mandrel and turned to the required shape. The periphery is covered with leather, coated with glue and rolled in emery until a considerable portion adheres to the glue-covered surface.

Wheels of this character will wear but a short time before the coating process must be repeated to form a new abrasive surface. They can scarcely be called grinding-wheels, and are more properly termed polishing wheels, and are used but very little except to produce a polished or finished surface. What are termed grinding-wheels, or "hard-wheels," are formed of emery in combination with some plastic mass that is preserved in moulds, in course of time becoming very hard like a grindstone.

If the mechanic desires a small grinding-wheel of this character, and cannot readily obtain one, he can make a very good substitute himself. To do this, procure a block of brass or cast iron, in which make a recess of the same diameter, but a little deeper than the desired thickness of the wheel. Make a hole centrally to the diameter of the recess and extending through the block, corresponding in size to the spindle on which the wheel is to be used. In this hole fit a strong bolt with one end threaded and a stout head on the other end. On the threaded end fit a nut. Make a thick washer that will fit pretty tight on the bolt, and at the same time fill the recess in the block. Make a follower of the same size that will fit in the same manner.

The materials for the wheels are glue and good emery. Make the glue thin, as for use on wood, and thicken with emery, and keep hot to be worked. When ready to make the wheels, oil the recess or mould as well as the washer and follower. This will prevent the hot mass from adhering to these parts. Put the washer at the bottom of the mould. Insert the bolt in the hole with the head at the bottom side of the block. Put in the hot glue and emery, well mixed together, spread it evenly in the mould, almost filling it. Put the follower on the bolt, letting it enter the mould and rest upon the glue and emery; then put on the nut and screw it down tight with a wrench. The mass is compressed according to the force employed.

If the wheel be small and thin it will cool and harden in a few minutes so that it can be removed. Take off the nut and follower and drive out the bolt, and if the recess be properly made a blow with a hammer on the bottom of the block will expel the wheel and washer.

In place of a recess cut in a block of metal, a ring may be used, care being taken to place it so that the bolt will be central, to insure equal radius on all sides.

Oiling the parts prevents the glue and emery from sticking. The washer put in the bottom of the mould facilitates the removal of the soft wheel, and also tends to prevent it from injury while being removed. The wheels must be dried in a warm place before being used, and must be kept away from moisture.

Above the size of two or three inches it would be hardly advisable to attempt making this kind of wheel.

Common shellac may be used in place of glue, but the objections to its employment are the greater cost, difficulty to mix with emery, and it is also more difficult to put in the mould. It has the advantage over the glue and emery wheel, inasmuch as it is proof against moisture or water. For a small, cheap wheel, and one that can be readily made, the one made of glue and emery is preferable. —*By* W. B.

MAKING AN EMERY WHEEL.

Having a few articles to polish I thought I would make an emery wheel. After turning my truck and fastening it to the arbor I tried several times to glue the leather to the truck or wheel and failed. The splice was what bothered me most. Looking around for a way out of this difficulty, I came across an old pair of woolen or felt boots such as are worn by loggers.

I took the leg of one of these boots, cut off a ring the width of my truck, glued it on the truck and turned it off as well as I could. I held a hot iron over it until it was very smooth, and then covered it with glue. I next heated emery as hot as I thought necessary, spread it on a board and rolled the truck in it and pounded it in. When it was dry I gave it another coat and then another. Three coats are enough, at least they were sufficient for the wheel I am using. A glance at the accompanying engraving, Fig. 226, will give anyone a fair idea of how the job should be done. —*By* H. A. Seavey.

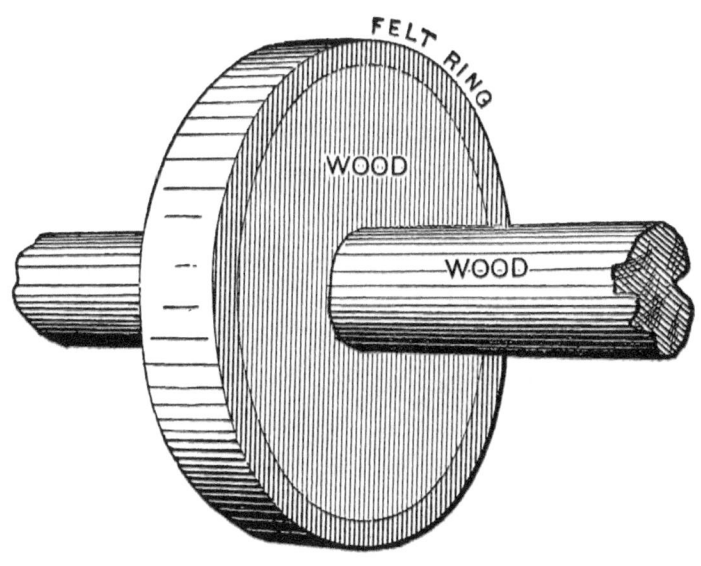

FIG. 226—MAKING AN EMERY WHEEL.

SOMETHING ABOUT GRINDSTONES AND GRINDING TOOLS.

In the matter of the average grindstone, its use and misuse, I would state that the result of my observation and experience is:

First—It is too small in diameter. Second—It is too broad-faced. Third—It is not properly speeded. Fourth—It is not properly cared for. Fifth— It is not properly used.

Stones should be narrow-faced to secure a greater proportion between that which is worn from its surface by useful work and that which is removed by the truing device. It is patent to every practical mechanic that the portion of a stone most in use is a very narrow line at each corner, and the reason for this is plain when we consider that after a tool is once properly shaped the workman will endeavor to confine his grinding to the top or cutting-face of the tool, leaving the sides and clearance angles intact, if possible, and to do this, keeping in mind the desired cutting-lip, he must have recourse to the corners to secure the proper inclination of the tool for that result. So it comes about that the corners are rapidly worn rounding.

It is a matter of experience that the faster a stone runs the faster it does its work and the longer it remains in working shape. But they are weak, and if run too rapidly, have an uncomfortable habit of disintegrating themselves. Water has to be used for the two-fold purpose of keeping the tools cool, and the stone clean and free from glaze, but water has a decided tendency to disassociate itself from a stone that capers around too lively. So we are compelled to reduce the speed to the fastest possible, compatible with safety and freedom from a shower bath. Now, of all the inconsistencies that exist in modern machine-shop practice, I think that the running of the average grindstone is the most pronounced, because it has not the adjunct of a variable speed due to the losses of diameter.

In regard to the choice, care and use of a stone, I would discourse as follows: The desiderata in the selection of a stone are, that it should cut fast, should not glaze, and should remain true. To secure the cutting and anti-glazing qualities—for they are associated—a stone should be close and sharp-grained, and not too firmly cemented or hard. It must be just soft enough to slowly abrade under the mark; such abrasion constantly brings new cutting points into prominence, and prevents the lodgment of the abraded particles of steel upon the stone, which would finally result in glazing. For the proper maintenance of its truth, it is essential that the stone be homogeneous, as uneven hardness must result in uneven wear. The condition of homogeneity is one that cannot exist in a natural stone, but ought reasonably to be expected in an artificial one, and I believe that the grindstone of the future will be manufactured—not quarried.

To get the best results from a stone filling the above requirements, it should be hung in a substantial frame, properly balanced, supplied with *clean* water, never allowed to stand immersed, because that softens locally and thus throws it out of balance. Therefore, I say that the average stone is not properly cared for and used, because these conditions for well-being are rarely met.

For ordinary tool-grinding, I would recommend that the "front" side of the stone be used, not because better work can be done there; but because it can usually be done there faster; and that it be fitted with an adjustable narrow-edged rest, used close to the stone, and extending around the sides toward the center about two inches. Such a rest enables one to incline his tool in any possible direction, and hold it firmly with adequate pressure, while running small risk of the dreaded "dig."

In all sorts of tool-grinding, my experience tells me that the cutting edge of a tool should always be toward the approaching side of the stone or wheel. — Ben Adriance, *in American Machinist*.

HANGING A GRINDSTONE.

To hang a grindstone on its axle so as to keep it from wabbling from side to side requires great skill. The hole should be at least three-eighths or one-half inch larger than the axle and both axle and hole square. Then make double wedges for each of the four sides of the square, all alike and thin enough so that one wedge from each side will reach clear through the hole. Drive the wedges from each side. If the hole through the stone is true the wedges will tighten the stone true. If the hole is not at right angles to the plane of the stone it must be made so or the wedge must be altered in the taper to meet the irregularity of the hole.

DEVICE FOR FASTENING A GRINDSTONE.

The device, a sketch of which I send you herewith, is very simple and effective for fastening a grindstone. The illustration, Fig. 227, shows the method so well that only a brief description appears to be necessary.

Almost any mechanic will see at a glance that the tightening of the screws or bolts (either can be used, according to the size of the grindstone), as shown in the engraving, cannot fail to hold the stone securely upon the shaft. —*By* H. G. S.

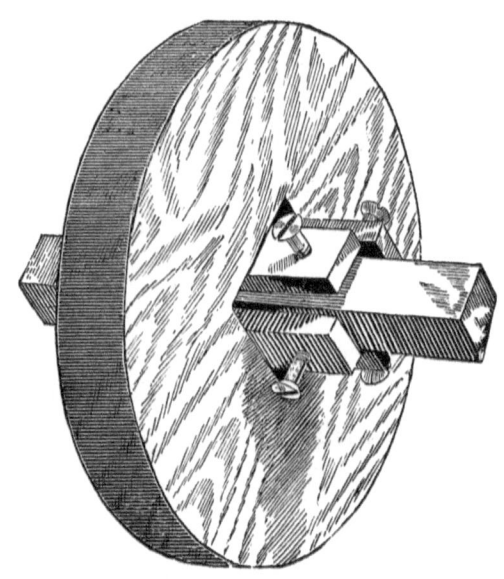

FIG. 227—DEVICE FOR FASTENING A GRINDSTONE, AS DESCRIBED BY "H. G. S."

Note. —It was not thought necessary to show the device of our correspondent attached to the grindstone frame, because that is a simple matter which probably every blacksmith or wheelwright understands thoroughly. The stone can be set true by loosening or tightening the opposite screws or bolts. —Ed.

MOUNTING A GRINDSTONE.

I send you some illustrations representing a convenient method of mounting a grindstone. The casting will cost about 50 cents for a 30-inch stone.

Anyone can make the pattern and core box. A in Fig. 228 of the engraving annexed, is a cast-iron flange, *bbbb* are set-screws tapped into the flange and impinging on the square bar, which is turned up with gudgeons, and will constitute the axis of the grindstone. Top holes corresponding to those marked C in the flange are drilled through the stone and a flange like A is bolted to each side of the stone.

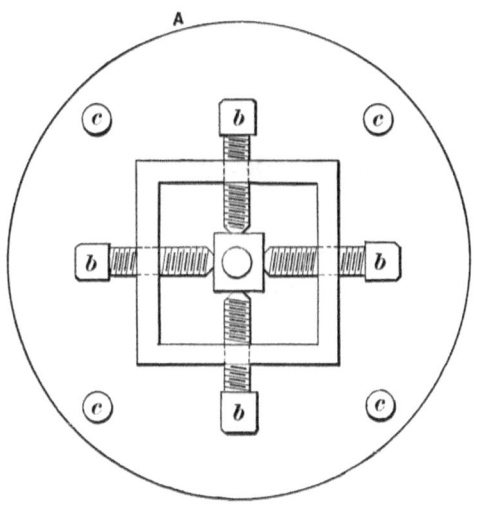

FIG. 228—MOUNTING A GRINDSTONE. END VIEW OF THE MOUNTING FLANGE.

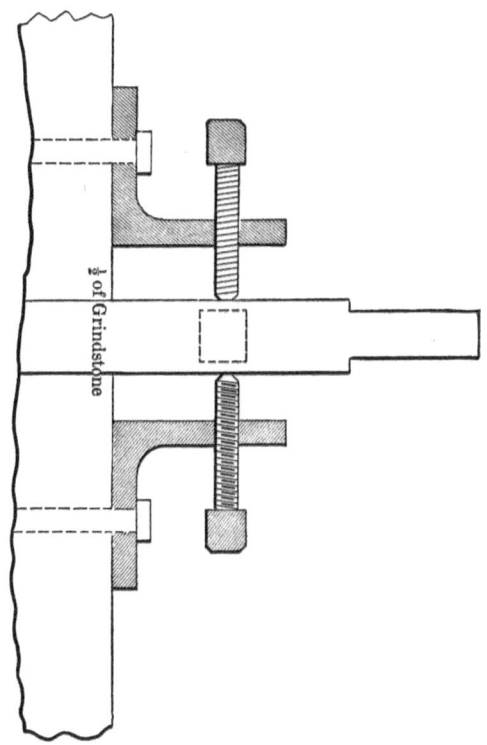

FIG. 229—SECTIONAL VIEW OF THE FLANGE.

Fig. 229 is a sectional view of the flange. By slacking and setting the screws b the stone can be made true to a hair. —*By* J. H. S.

HOW TO MAKE A POLISHING MACHINE.

An emery wheel is a good thing for every smith to have about his shop, for burnishing axles, chisels, knives, and a thousand things that should be kept bright. But owing to the high price of wheels, together with the fact that few shops have any steam or water power to run them, but very few are used.

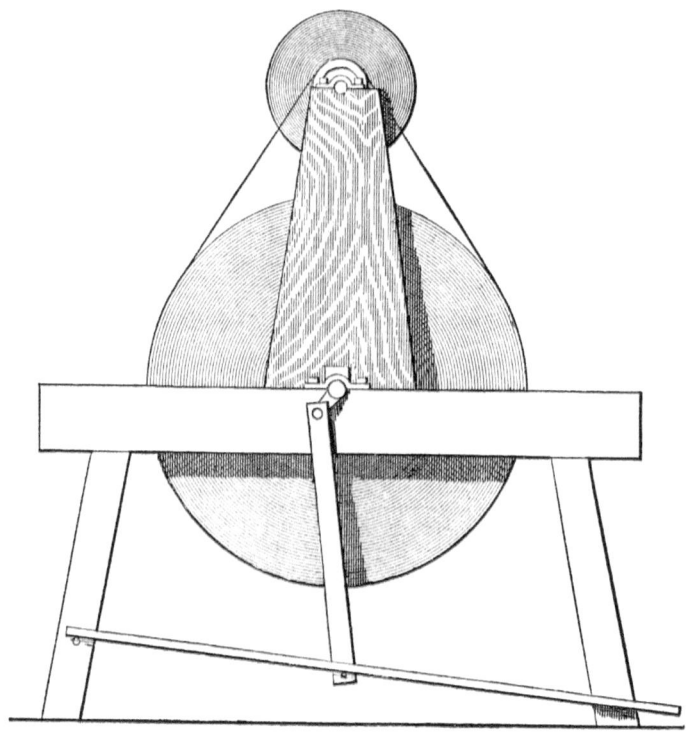

FIG. 230—A POLISHING MACHINE.

Now any smith can get up a rig for polishing that will answer every purpose and cost very little. First make a driving wheel, as shown in the accompanying illustration, Fig. 230, 2 ½ feet in diameter by 4 to 6 inches in thickness. It can be made with arms, or solid, by nailing together several thicknesses of boards.

Hang it in a frame as you would a grindstone. Put upon one end of the crank a balance wheel of not less than 50 lbs. weight and attach the other end to the foot piece by a rod; or, what is better, a piece of hard wood, which will not wear on the crank. Bolt two pieces of hardwood board, 4 x 8 and 1 inch in thickness, upon the inside of the frame. Cut a slot at the wide end to go astride of the crank; also, one at the narrow end to receive the spindle of the emery wheel. The emery wheel should be some 10 or 12 inches in diameter, by 3 inches in thickness, with a small pulley bolted upon one side for the belt. Now, put on the belt, apply the foot power, and turn the emery wheel as true as possible with a sharp tool. Then cover it with sole leather. The leather should be well soaked in hot water and pegged on wet; one row of pegs at the edges, one inch apart, will be sufficient. After the leather is put on, fix a rest and true it again as before.

To emery the wheel, make a box 2 ½ feet long and a little wider than the wheel, in which to put the emery. After putting on a coat of thin glue with a brush, roll the wheel in the box and the coating is done at once. It should stand a few hours before using. Whenever the wheel gets smooth and doesn't cut, apply another coating as before. After several coatings have been put on the old emery should be removed, which can be done by soaking in hot water and scraping with a knife or piece of glass. It is a good thing to have two wheels, one for coarse and the other for fine emery. I have a rig which I made in this way ten years ago and it works like a charm. The expense of running it is next to nothing: try it and you will not be without one. —*By* W. H. Ball.

END OF VOL. II.

www.ingramcontent.com/pod-product-compliance
Lightning Source LLC
Chambersburg PA
CBHW020421010526
44118CB00010B/364